EARTH SCIENCES IN THE 21ST CENTURY

THREE-DIMENSIONAL MICRO-METEOROLOGICAL MODELING IN URBAN AREAS

SURFACE ENERGY BALANCE AND POLLUTANT DISPERSION

EARTH SCIENCES IN THE 21ST CENTURY

Additional books in this series can be found on Nova's website
under the Series tab.

Additional e-books in this series can be found on Nova's website
under the e-books tab.

EARTH SCIENCES IN THE 21ST CENTURY

THREE-DIMENSIONAL MICRO-METEOROLOGICAL MODELING IN URBAN AREAS

SURFACE ENERGY BALANCE AND POLLUTANT DISPERSION

YONGFENG QU

nova publishers
New York

Library of Congress Cataloging-in-Publication Data

ISBN: 978-1-63483-757-6

Published by Nova Science Publishers, Inc. † New York

Contents

Preface

Local atmospheric simulations of real city neighborhoods are important to study for urban climate, thermal and wind comfort, and local air quality, including pollution hot spots such as tunnel heads, near a motorway or a busy crossing. When the wind is weak, the thermal and radiative effects become more important together with flow stratification that can inhibit turbulence and mixing in some conditions. In order to take into account atmospheric radiation and the thermal effects of the buildings in simulations of atmospheric flow and pollutant dispersion in urban areas, a three-dimensional atmospheric thermal-radiative scheme in the atmospheric module of the open-source Computational Fluid Dynamics (CFD) model has been developed. The simulation of realistic atmospheric conditions in urban areas made possible by this work can be used for various applications. The author describes the development of this model and presents several results by comparing field data.

PACS: 05.45-a, 52.35.Mw, 96.50.Fm

Keywords: surface energy balance, atmospheric dispersion, computational fluid dynamics, thermal transfer

Acknowledgments

Modeling an urban surface energy system is a multi-disciplinary work and the fruit of the collaboration of international teams. Before going into technical details, I would like to pay tributes to and name all my colleagues, both French and foreign, who have accompanied me these years, colleagues whose work and ideas have left a mark which one discovers between the lines of the program: Maya Milliez, Bertrand Carissimo, Luc Musson-Genon, Patrice Mestayer, Majorie Musy, Valéry Masson, Jean-François Sini, Jean Philippe Gastellu-Etchegorry, Marina Neophytou, Richard Howard, Alexandre Douce, Namane Méchitoua, James Voogt, Jörg Franke, Bert Blocken, Dominique Groleau, Christian Seigneur, Denis Wendum, Raphael Bresson, Hanane Zaïdi, Julien Bouyer, Laurent Malys, Noëlie Daviau, Benoit Gauvreau, Emilie Redon, George Fitton, Gauthier Soleilhac.

I am also grateful to Meteo-France for providing access to the CAPITOUL data, the Defense Threat Reduction Agency (DTRA) for providing access to the MUST data and the French Research Agency for sponsoring the EUREQUA project.

Nomenclature

Superscripts

$\downarrow$	incoming	
$\uparrow$	outgoing	-
n	timestep n	-

Subscripts

int	internal building or deep soil	
ref	reference value	-

Roman Symbols

B_p, B_d	drag-porosity model constant	-
C	heat capacity	$J\ K^{-1}\ m^{-2}$
CFL	Courant-Friedrichs-Lewy number	-
C_D	drag coefficient	-
$C_{4\varepsilon}, C_{5\varepsilon}$	drag-porosity model constant	-
C_p	specific heat capacity	$J\ kg^{-1}\ K^{-1}$
C_μ	turbulence model coefficient	-
d	distance of the cell center to the wall	m
D_t	turbulent diffusivity	$m^2\ s^{-1}$
e	wall thickness	m
e_u	water vapor pressure	Pa
E_1	constant in the Bird model associated with aerosol absorption	-
f_m, f_h	Louis stability functions	-
F	cloud factor	-
F_c	ratio of the forward-scattered irradiance to the total scattered irradiance due to aerosols	-
F_i	drag force	N
g	gravity	$m\ s^{-2}$
$\widetilde{G}$	production or destruction rate due to buoyancy	-
h	altitude angle of the sun	rad
h_f	heat transfer coefficient	$W\ m^{-2} K^{-1}$
H	height	m
$I(\underline{x}, \underline{D})$	intensity of radiation at the location ($\underline{x}$) along the propagation direction with the vector $\underline{D}$	$W\ m^{-3}\ sr^{-1}$

$I(z)$	turbulent intensity	-
k_s	equivalent roughness height	m
$k(z)$	turbulent energy	$m^2\ s^{-2}$
K	surface conductance	$W\ K^{-1}\ m^{-2}$
$K_a(\underline{x})$	radiation absorption coefficient	-
L	long-wave radiative flux	$W\ m^{-2}$
L_a	long-wave radiative flux from the sky	$W\ m^{-2}$
L_e	long-wave radiative flux from the multi-reflection on the other surfaces	$W\ m^{-2}$
$L(\underline{x},\underline{D})$	monochromatic luminance	$W\ m^{-3}\ sr^{-1}$
m	air mass	$kg\ m^{-3}$
p	pressure	Pa
Pr_t	turbulent Prandtl number	-
$\widetilde{P}$	production rate of turbulent energy	-
q_{rad}	radiative heating rate	$W\ m^{-3}$
Q	release rate	$kg\ s^{-1}$
Q_{cond}	conductive heat flux	$W\ m^{-2}$
Q_E	latent heat flux	$W\ m^{-2}$
Q_F	anthropogenic heat flux	$W\ m^{-2}$
Q_H	sensible heat flux	$W\ m^{-2}$
ΔQ_A	advective heat flux	$W\ m^{-2}$
ΔQ_S	storage heat flux	$W\ m^{-2}$
Q^*	net radiation flux	$W\ m^{-2}$
Re_L	Reynolds number based on the integral scale of the flow	-
S	short-wave radiative flux	$W\ m^{-2}$
S_0	solar constant (= 1370)	$W\ m^{-2}$
S_{dr}	solar Rayleigh diffusion flux	$W\ m^{-2}$
S_{dae}	solar aerosol diffusion flux	$W\ m^{-2}$
S_{dg}	solar reflection flux by the surface	$W\ m^{-2}$
S_D	direct solar flux	$W\ m^{-2}$
S_{Dn}	direct normal solar flux	$W\ m^{-2}$
S_e	solar flux diffused by the environment	$W\ m^{-2}$
S_f	solar flux diffused by the atmosphere above	$W\ m^{-2}$
S_k, S_ε	extra source terms of turbulence and dissipation	-
S_{rad}	extra thermal source	$J\ K\ m^{-5}s^{-1}$
T	temperature	K, °C
T_a	air temperature	K, °C
T_{br}	brightness surface temperature	K, °C
$T_{g/b}$	deep soil or internal building temperature	K, °C
T_{se}, T_{sfc}	external surface temperature	K, °C
u_*	friction velocity	$m\ s^{-1}$
U	wind speed	$m\ s^{-1}$
v_t	turbulent viscosity	$m^2\ s^{-1}$
$\widetilde{v}$, v', $\overline{v}$	Favre average for RANS, its fluctuation,	

	the ensemble average	-
w	precipitation water content	-
x	distance from the emission source	m
y_P	a distance from the center point P of the wall-adjacent cell to the wall	m
z_g	ground level	m
z_G	boundary layer height	m
z_0	aerodynamic roughness length	m
z_{0_T}	thermal roughness length	m
z^+	dimensionless wall distance z/z_0	-
Z	sun zenith angle	rad

Greek Symbols

$\alpha, \alpha_g, \alpha_s$	albedo, averaged ground albedo, clear sky or atmospheric albedo	-
β	wind power-law exponent	-
ε	emissivity of the surface	-
ε_{clr}	effective or apparent emissivity	-
$\varepsilon(z)$	dissipation rate	-
κ	von Kármán constant	-
λ	average thermal conductivity	$W\,m^{-1}K^{-1}$
μ	viscosity of the fluid	$kg\,m^{-1}s^{-1}$
μ_t	thermal admittance	$J\,m^{-2}\,s^{-0.5}\,K^{-1}$
ν	kinematic viscosity	$m^2\,s^{-1}$
ω	Earth angular frequency	Hz
$d\Omega$	element of the solid angle (sr) around the direction $\underline{D}$	-
ϕ	wind angle	$^\circ$
ρ	density	$kg\,m^{-3}$
σ	Stefan-Boltzmann constant ($= 5.66703 \times 10^{-8}$)	$W\,m^{-2}\,K^{-4}$
σ_t	turbulent Prandtl number	-
σ_y, σ_z	horizontal plume width, vertical plume height	m
τ	number of seconds in a day	s
τ_w	wall shear stress	$N\,m^{-2}$
τ_a	transmittance due to aerosol absorption and scattering	-
τ_{aa}	transmittance due to aerosol absorption	-
τ_{ad}	transmittance due to aerosol scattering	-
τ_{aw}	transmittance due to water vapor absorption	-
τ_g	transmittance due to mixed gas absorption	-
τ_{oz}	transmittance due to ozone absorption	-
τ_r	transmittance due to Rayleigh diffusion	-
θ	potential temperature	K, °C

Acronyms

1D, 2D, 3D	one-, two-, three-dimensional
ABL	Atmospheric Boundary Layer
AIJ	Architectural Institute of Japan
AMS	American Meteorological Society
CAD	Computer-Aided Design
CAPITOUL	Canopy and Aerosol Particles Interactions in TOulouse Urban Layer field experiment
CEREA	Centre d'Enseignement et de Recherche en Environnement Atmosphérique (Atmospheric environment teaching and research center)
CERMA	CEntre de Recherche Méthodologique d'Architecture (Center for Methodological Research in Architecture
CFD	Computational Fluid Dynamics
Code_Saturne	3D open source CFD code developed by EDF and CEREA
COST	European COoperation in Science and Technology research group
DART	Discrete Anisotropic Radiative Transfer model
DNS	Direct Numerical Simulation
DOM	Discrete Ordinates Method
DTRA	Defense Threat Reduction Agency
EDF	Électricité De France (French electricity company)
ERCOFTAC	European Research Community on Fluids, Turbulence And Combustion organizations
EUREQUA	Evaluation mUltidisciplinaire et Requalification Environnementale des Quartiers (Multidisciplinary Assessment and Environmental Requalification of neighborhoods project)
FOV	angular Fields Of View
IOP	Intensive Observation Periods
IPCC	Intergovernmental Panel on Climate Change
IRT	InfraRed Thermometer
JAPEX	Joint ATREUS-PICADA EXperiment
LES	Large Eddy Simulation
LST	Local Standard Time
LUMPS	Local-scale Urban Meteoro-logical Parameterization Scheme
MBE	Mean Bias Error
MUST	Mock Urban Setting Test field experiment
NIWA	National Institute of Water and Atmospheric Research, New Zealand
RANS	Reynolds-Averaged Navier-Stokes
RMSE	Root Mean Square Error
RNG	ReNormalization Group
RTE	Radiative Transfer Equation
SEB	Surface Energy Balance
SOLENE	thermo-radiative software developed by CERMA
SUMM	Simple Urban Energy Balance Model for Mesoscale Simulations

SVF	Sky View Factor
SYRTHES	SYstèm de Résolution THErmique Solide code developed by EDF
TEB	Town Energy Balance scheme
TKE	Turbulent Kinetic Energy
TIR	Thermal InfraRed
TOPEUM	Towards Optimisation of urban-planning and architectural Parameters for Energy Use minimisation in Mediterranean cities campaign¿
TUF-3D	Temperatures of Urban Facets in 3D
PBL	Planetary Boundary Layer
UBL	Urban Boundary Layer
UCL	Urban Canopy Layer
UCLCM	Urban Canopy Layer Climate Model
UHI	Urban Heat Island
UT	Universal Time
WBCSD	World Business Council for Sustainable Development
WHO	World Health Organization

Chapter 1

Introduction

1. Background

Urbanization is modernization, industrialization and mobilization. It is a transformation in social environment, political organization and division of work. Rapid urban extension in the last century has had far-reaching consequences for sustainability of the environment and had profoundly changed the environment by many activities. More than half of the world's population lives in urban areas. Oke (1987) describes how the process of urbanization produces radical changes in the nature of the surface and atmospheric properties of a region. Urbanization changes the radiative, thermal, hydrologic and aerodynamic characteristics of the atmosphere and thereby alters the natural energy and hydrology balances, as well as the wind and turbulence levels.

Therefore, among the many issues and challenges involved in urban development, are the environmental ones, such as:

- Urban heat island
- Air pollution dispersion
- Pedestrian wind comfort
- Energy management

1.1. Urban Heat Island

The Urban Heat Island (UHI) phenomenon is an increase in air temperature within cities compared to the city's rural surroundings. It was first identified by Howard (1820) over the city of London. Especially at night, the air temperature difference can be up to 3 to 10 K for large agglomerations (Oke, 1987). The deterioration of the urban thermal environment has been recognized as a serious problem during the summer months even in mid-latitude regions. During heat waves, it can lead to serious consequences in terms of public health. An example of this was revealed during an important heat wave in August 2003 that affected Europe and caused more than 70, 000 victims, with a majority in urban areas (Hémon and Jougla, 2003). This kind of phenomenon could become worse and more frequent if climate change predictions are accurate (temperature increases even more in cities due to a positive feedback) (McCarthy et al., 2010).

Consequently, urban development initiatives that consider influences on urban thermal environment have received more attention than they have in the past. In addition, a thermally comfortable environment would be pleasant for inhabitants and commuters in urban areas.

It is also worth noting that the UHI is seen during both summer and winter months. Pragmatic use of the UHI effect in winter may prove advantageous: reducing the need for heating, making the snow on the roads melt faster etc. Moreover, most plants are sensitive to temperatures and only grow above a certain threshold. In areas which are affected by the UHI effect there is more growth in most plants, so it may be practical from an agricultural point of view.

1.2. Air Pollution Dispersion

In urban environment, especially near busy traffic, the population is exposed to high levels of hazardous material concentrations, exceeding current air quality standards. The combination of high population density and high traffic volume in urban areas means that the health impact of traffic pollution can potentially be much larger than similar sources in less populated areas. Street canyons can exacerbate the health impact of traffic pollutants, hence the need to understand their dispersion dynamics.

The heat wave that struck Europe in summer 2003 was not only extreme in temperature but also in the persistence of high ozone concentrations for almost three weeks. The World Health Organization (WHO) states that more than two million people die each year from causes directly attributable to air pollution (WHO, 2006). Current research concludes that emissions in buildings are one of the major sources of the pollution that causes urban air quality problems, and pollutants that contribute to climate change. They account for 49% of sulfur dioxide emissions, 25% of nitrous oxide emissions, and 10% of particulate emissions, all of which damage urban air quality. From the source of World Resource Institute, based on data for 2000, buildings (including residential and commercial buildings) produce 15.3% greenhouse gas emissions ahead of industry (10.4%) and transportation (13.5%) sectors. In that event, sustainable development requires the improvement of the interrelationships between a building, its components, its surroundings, and its occupants.

1.3. Energy Management

The World Business Council for Sustainable Development (WBCSD, 2009) points out that 40% of the world's energy is consumed in buildings. New buildings that will use more energy than necessary are being built every day, and millions of today's inefficient buildings will remain in 2050. On one side, in order to reduce the global energy-related carbon footprint by 77% or 48 Gigatons to stabilize CO_2 levels in order to reach the ones recommended by the Intergovernmental Panel on Climate Change (IPCC), the building sector must radically make cut in energy consumption. On the other side, this presents an excellent opportunity for business to develop new products and services that cost-effectively reduce the energy burden on consumers and countries while contributing to the slowdown of climate change. This market could be worth between US$ 0.9 trillion and US$ 1.3 trillion.

1.4. Pedestrian Wind Comfort

Near high-rise buildings, winds may reach high velocities at pedestrian levels, contributing to the general discomfort of the city inhabitants or even being dangerous. In March 2011, in the United Kingdom, high winds blustered in Yorkshire for almost one day long, causing some minor structural damages to buildings and roads. Moreover, a report of the Daily Mail described how a pedestrian was killed and another injured when a lorry overturned and toppled over them during a high wind episode in the center of the city of Leeds. Actually, there have already been recorded instances of people being blown off their feet near high-rise buildings. Lawson and Penwarden (1975) reported dangerous wind conditions to be responsible for the death of two old ladies in 1972 after being blown over by sudden wind gusts.

On the other hand, in weak wind regions, good urban ventilation helps to reduce the UHI. This is definitely an advantage with the raising concerns regarding the cost and environmental impact of energy use. Not only does the wind provide natural ventilation (outdoor air) to ensure safe healthy and comfortable conditions for building occupants without the use of fans, it also provides free cooling without the use of mechanical systems. Adequate urban ventilation is also helpful to increase air pollutant dispersion around buildings (Shirasawa et al., 2008).

In fact, the construction of a building inevitably changes the microclimate and the ventilation in its vicinity. Therefore, the design of a building should not only focus on the building appearance and on providing a good indoor environment but should also include the effect of its architecture on the outdoor environment. The impact of buildings on the outdoor environment, in particular, related to the wind, has received relatively little attention, so far.

2. Urban Physics

As a consequence of these issues, focus was put on the research of urban physics, aiming to better understand and model the phenomena occurring in urban areas and the atmosphere above, such as heat, moisture and momentum transfers, pollutant and acoustic dispersion, radiative transfer etc. Urban physics cover a large range of disciplines: meteorology, fluid dynamics, thermal, aerodynamics, and acoustics etc. For instance, in the field of wind engineering, urban physics analyze the effects of the wind in the built-up environment and studies the possible damages or benefits which may result from the wind. In the fields of air pollution, urban physics also includes low and moderate winds as these are relevant to dispersion of contaminants.

2.1. Spatial and Temporal Scales and Urban Boundary Layer

The interactions between urban areas and the atmosphere above imply various scales. Orlanski (1975) gives a rational subdivision of scales for atmospheric processes:

- synoptic scale (scale of largest cyclones, distance larger than 2000 *km*)

- mesoscale (between synoptic scale and larger than microscale, from 2000 *km* to 200 *km* and from 200 *km* to 2 *km*)
- microscale (near-ground atmospheric phenomena, the distance less than 2 *km*). More recently, microscale includes smaller scales, such as: building scale (less than 100 *m*), building component scale (less than 10 *m*) even building material scale.

Scale-dependent parameterizations are needed to include the influences of built-up areas on meteorological fields in atmospheric models. In order to study the impact of an urban area on its surroundings (scale varying from 10 to 100 *km*) mesoscale modeling is used, where the influence of obstacles is parameterized. In order to study the phenomena occurring within the urban canopy (scale varying from 100 *m* to 2 *km*) obstacle resolving microscale models are used, where the obstacles are explicitly included in the mesh.

While characteristic scales of phenomena resulting from single obstacles are relatively small ($\sim 100\ m$, $\sim minutes$) and can be resolved, in reality there are always multiple obstacles, as for example in urban areas; many buildings, in a wind turbine park; many turbines, or in a forest; many trees. In these cases, only a few of the obstacles can be resolved with sufficient detail, and their impacts need to be parameterized (Schlünzen et al., 2011). To calculate the interaction of canopy layer processes and the air above, a coupling between mesoscale and microscale models can be used. Orlanski (1975), Randerson (1976) and Schlünzen et al. (2011) summarize the spatial scales of phenomena that can be directly simulated in a mesoscale and microscale model, detailing the scales that have to be parameterized in a model or to be considered via the boundary values.

The boundary layer over an urban area is of particular interest as it is in this layer of the atmosphere that the majority of observations in urban areas are made (Oke, 1987; Stull, 1988). It is, therefore, important to know what these observations represent. As air flows from one surface to another an internal boundary layer forms. The internal boundary layer is influenced by the new surface and deepens with fetch. The internal boundary layer formed over urban areas is the Urban Boundary Layer (UBL) (Stull, 1988; Oke et al., 1999). The buildings introduce a large amount of vertical surfaces, high roughness elements, artificial materials, and impervious surfaces (such as buildings and pavements that are made of dark colors that absorb the heat which causes the temperature of the surface and surrounding air to increase). The most well-known consequences are the UHI, the generation of local flows between the city center and its outskirts and between the various city districts, and the 'urban plume' (a dome of heated air above the cities due to pressure differences between warmer temperatures in the city and cooler temperatures in the surrounding rural areas) downwind of a city. In calm or low wind conditions, the warmer air in the city core rises, pulling air near the surface radially inward and a radially outward return flow develops aloft. This air circulation forms the 'urban dome'

2.2. Urban Surface Energy Balance

Knowledge of the surface energy balance is essential to understand urban climate and boundary layer processes. The energy balance containing different heat fluxes can be defined as illustrated in Figure 1.1, with the following equilibrium relationship:

$$S^* + L^* + Q_F = Q_H + Q_E + \Delta Q_S + \Delta Q_A, \tag{1.1}$$

where S^*, L^* ($W\ m^{-2}$) are the net short-wave and long-wave radiation flux respectively, Q_F ($W\ m^{-2}$) is the anthropogenic heat flux release within the building-air system, Q_H ($W\ m^{-2}$) and Q_E ($W\ m^{-2}$) refer to convection of sensible heat flux and latent heat flux respectively. The terms ΔQ_S ($W\ m^{-2}$) and ΔQ_A ($W\ m^{-2}$) are storage of heat in the ground and the buildings and advective heat transfer within the building-air system. Note that here the terms 'convection' and 'advection' refer to vertical turbulent transfer and mean horizontal transfer, respectively.

For dry surfaces, the energy balance in equation (1.1) is simplified by neglecting Q_E, and if the sites are horizontally homogeneous, ΔQ_A can be also ignored. Therefore, surface temperature results from the balance of energy exchanges at the surface given the incoming radiative forcing, the local ambient air temperature, and the surface radiative properties.

Radiative Balance Q^*

The net radiative flux at a surface reads:

$$Q^* = S^{\downarrow} - S^{\uparrow} + L^{\downarrow} - L^{\uparrow}, \tag{1.2}$$

where $S^{\downarrow}$ ($W\ m^{-2}$) and $S^{\uparrow}$ ($W\ m^{-2}$) are respectively the incoming and outgoing short-wave radiative fluxes, $L^{\downarrow}$ and $L^{\uparrow}$ are respectively incoming and outgoing long-wave radiative fluxes ($W\ m^{-2}$).

Incoming short-wave radiation can be decomposed into its direct and diffuse component, and for both short- and long-wave radiation we can distinguish the contributions coming from the atmosphere above and from the urban environment:

$$S^{\downarrow} = S_D + S_f + S_e, \tag{1.3}$$

$$L^{\downarrow} = L_a + L_e, \tag{1.4}$$

where S_D ($W\ m^{-2}$) is the direct solar flux, S_f ($W\ m^{-2}$) the solar flux diffused by the atmosphere above, S_e ($W\ m^{-2}$) the flux diffused by the environment (i.e., from multi-reflection on the surfaces). L_a ($W\ m^{-2}$) and L_e ($W\ m^{-2}$) are the long-wave radiation fluxes from the sky and from the multi-reflection on the other surfaces.

Outgoing solar radiation expressed by the albedo (usually written α, a dimensionless quantity) which is the fraction of solar radiation reflected by a surface. Albedo determines how much solar energy, a particular substance reflects. Hence, $S^{\uparrow}$ reads:

$$S^{\uparrow} = \alpha(S_D + S_f + S_e). \tag{1.5}$$

Outgoing long-wave radiation is decomposed into an emitted and reflected part. It is a function of surface temperature T_{sfc} (K) and surface emissivity (usually written ε, a dimensionless quantity). Emissivity of a particular material is the fraction of energy that would be radiated by a black body at the same temperature. For a black body, ε would be equal to 1, while for any real object $0 < \varepsilon < 1$. Thus, $L^{\uparrow}$ reads:

$$L^{\uparrow} = \varepsilon\sigma T_{sfc}^4 + (1-\varepsilon)(L_a + L_e), \tag{1.6}$$

where σ is the Stefan-Boltzmann constant ($5.66703 \times 10^{-8}\ W\ m^{-2}\ K^{-4}$).

Anthropogenic Heat Flux Q_F

Anthropogenic heat flux is generated by humans and human activity, it has a small influence on rural temperatures, and so becomes more significant in dense urban areas (Washington, 1972). The American Meteorological Society (AMS) defines it as: "Heat released to the atmosphere as a result of human activities, often involving combustion of fuels. Sources include industrial plants, space heating and cooling, human metabolism, and vehicle exhausts. In cities, this source typically contributes $15 \sim 50\ W\ m^{-2}$ to the local heat balance, and several hundred $W\ m^{-2}$ in the center of large cities in cold climates and industrial areas."

Anthropogenic heat flux is one contributor to UHI. Although it is usually smaller (not in high-latitude cities) compared to other fluxes, its influence is observable (Pigeon et al., 2007). Anthropogenic heat generation can be estimated by adding all the energy used for heating and cooling, running appliances, transportation, industrial processes, plus that directly emitted by human metabolism.

Sensible Heat Flux Q_H

Surface sensible heat flux is the energy exchanged between a surface and the air in the presence of a surface-air thermal gradient. Modeling the sensible heat flux contributes to determine both stratification effects on turbulent transport, and to estimate the surface temperature. The sensible heat flux Q_H can be parameterized as:

$$Q_H = h_f(T_a - T_{sfc}), \tag{1.7}$$

in which h_f ($W\ m^{-2}K^{-1}$) is the heat transfer coefficient and T_a (K) the air temperature.

Storage Heat Flux ΔQ_S

The storage heat flux is a significant component of the energy balance. Knowledge of the storage heat flux term is required in a variety of applications, for example, modeling evapotranspiration, sensible heat flux, boundary layer growth, etc. Furthermore, the thermal inertia provided by this storage term is often regarded as a key process in UHI. It accounts for 17% to 58% of the daytime net radiation, and is greater at the more urbanized sites (downtown and light industrial). Considered at the hourly timescale ΔQ_S is variable. However, ΔQ_S is difficult to measure or model because of the complex three-dimensional (3D) structure of the urban surface and the diversity of materials. It is often determined as the residual of the surface energy balance equation (Grimmond and Oke, 1999b).

Camuffo and Bernardi (1982), Grimmond and Oke (1999b) suggest a hysteresis-type equation to characterize the storage heat flux as a linear function of Q^* and of the temporal variation of Q^*:

$$\Delta Q_S = a_1 Q^* + a_2 \frac{\partial Q^*}{\partial t} + a_3, \tag{1.8}$$

where t (s) is time. The parameter a_1 describes the overall strength of the dependence of the storage heat flux on net radiation. The parameter a_2 is the coefficient of retardation of ΔQ_S with respect to Q^*. The parameter a_3 is an intercept term that indicates the relative time when ΔQ_S and Q^* turn negative. Parameters a_1, a_2 and a_3 can be calculated through regression for hourly averaged data.

3. The Contents of This Monograph

This work aims to contribute to the study of detailed energy exchanges between buildings and the urban atmosphere (the distribution of surface temperatures), and to investigate pollutant dispersion around buildings and in urban areas. It involves developing a model coupling thermal transfer involving the buildings and a Computational Fluid Dynamics (CFD) modeling of the atmosphere in urban areas and discusses the findings to give insight into applications. The numerical model used is the atmospheric module of the 3D open source CFD code *Code_Saturne* developed by EDF and CEREA.

Chapter 2 presents some basic CFD aspects and urban energy models, then describes the latest development in the thermal-radiative-dynamical coupled model. Chapter 3 examines the model under academic conditions including a validation of the thermal-radiative-dynamical model with observations, a comparison between the radiative model with the SOLENE model and a numerical investigation on the thermal impact on a low wind speed airflow within an idealized built-up area. Chapter 4 performs the numerical simulation of two real urban area cases including data analysis, complex mesh generation, and simulation results. In Chapter 5, details are given on passive pollutant dispersion from complex urban layout under different thermal conditions for the street and the buildings using the developed model. Finally, Chapter 6 highlights the main conclusions.

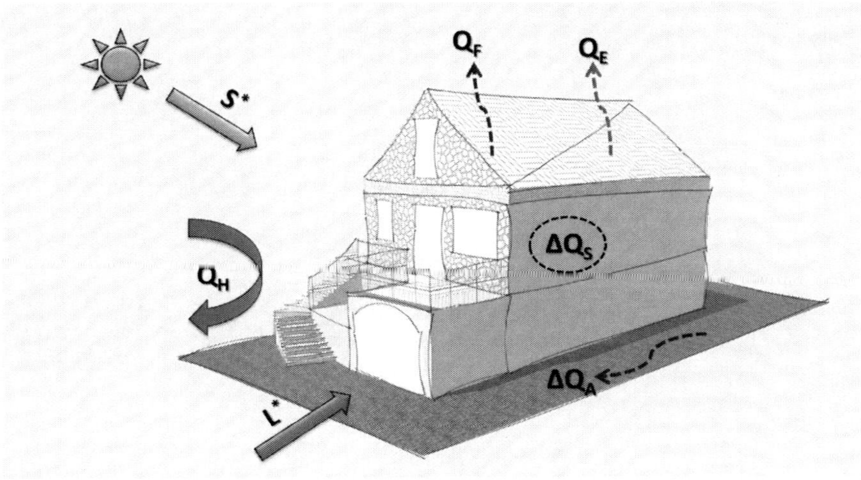

Figure 1.1. Schematics of the urban energy balance, adapted after Oke (1987). With: S^* $(W\ m^{-2})$ the net short-wave radiation; L^* $(W\ m^{-2})$ the net long-wave radiation; Q_H $(W\ m^{-2})$ the sensible heat flux; Q_E $(W\ m^{-2})$ the latent heat flux; ΔQ_S $(W\ m^{-2})$ storage; ΔQ_A $(W\ m^{-2})$ the advective flux; and Q_F $(W\ m^{-2})$ the anthropogenic heat flux.

Chapter 2

Model Design

1. General CFD Modeling Approach for the Urban Environment

In the past decades, Computational Fluid Dynamics has been intensively used to evaluate the indoor environment of buildings and heat and mass transfer between the indoor environment and the building envelope (Loomans et al., 2008). It has also been used in research on wind flow and the related processes in the outdoor environment around buildings, including pedestrian wind comfort (Blocken, 2009), wind-driven rain on building frontage (Briggen et al., 2009), pollutant dispersion (Hanna et al., 2006), exterior surface heat transfer (Blocken et al., 2009), natural ventilation and wind load of buildings (Cook et al., 2003). For both indoor and outdoor environment studies, the advances in computing performance and the development of efficient and powerful grid generation techniques and numerical solvers have led to the present situation in which CFD can technically be applied to study cases involving complex geometries and complex flow fields.

Numerical CFD modeling offers considerable advantages because it allows investigation where experimentation is not possible. It can provide a large amount of detail about a flow in the whole calculation domain, under varied conditions and without similarity constraints. The main limitations are the requirement of systematic and CFD solution verification and validation studies. The Navier-Stokes equations are commonly used to model the flow in the Atmospheric Boundary Layer (ABL) and the nature of the flow in an urban area, consisting of an arbitrary aggregation of buildings, is dominated by unsteady turbulent structures. Unfortunately, turbulent flow is one of the unsolved problems of classical physics. Despite many years of intensive research, a complete understanding of turbulent flow has not yet been attained (Davidson, 2004).

For turbulent flows, the range of length scales and complexity of phenomena involved in turbulence make most modeling approaches prohibitively expensive; the resolution required to resolve all scales involved in turbulence is beyond what is computationally possible. The primary approach in such cases is to create numerical models to approximate unresolved phenomena. In addition to the wide range of length and time scales and the associated computational cost, the governing equations of fluid dynamics contain a non-linear convection term and a non-linear and non-local pressure gradient term. These nonlinear equations must be solved numerically with the appropriate boundary and initial conditions. Several methods exist for predicting turbulent flows with CFD. Three of the most popular approaches are: Direct Numerical Simulation (DNS), Large Eddy Simulation (LES) and Reynolds-Averaged Navier-Stokes (RANS) simulation. Detailed introduction to these methods are beyond the scope of this monograph. Here, I just briefly present the advantages and weaknesses of each simulation approach.

Direct Numerical Simulation (DNS). DNS solves the exact Navier-Stokes equations by resolving all the scales of motion from the energetic large scales to the dissipative small scales, without any modeling. In consequence, DNS is expected to provide accurate predictions of the flow (Moin and Mahesh, 1998). However, the associated computational cost is extremely expensive in the case of urban flow problems. Indeed, the number of grid points required to simulate a 3D turbulent flow in DNS is proportional to $Re_L^{9/4}$, where Re_L is the Reynolds number based on the integral scale of the flow. Because the time step is related to the grid size, the total computational cost for DNS actually increases as Re_L^3. This rapid increase in Re_L prohibits the application of DNS to high Reynolds number flows, such as the ones in the ABL. Despite all of the progress in terms of computational power, DNS is still restricted to flows with low Reynolds numbers with relatively simple obstacles in urban areas because of the very large range of scales that have to be resolved (Coceal et al., 2007).

Large Eddy Simulation (LES). The basic idea of LES is to solve 'filtered' Navier-Stokes equations, therefore to resolve only the large-scale motions in a turbulent flow and model the small-scale (unresolved) motions. The latter scales of motion are expected to be more universal and, hence, easier to model. Compared to DNS, LES is not an exact solution but is less computationally demanding. However, the application of LES to wall-bounded flows, particularly at high Reynolds numbers, is severely restricted owing to the grid resolution requirements for LES to resolve the viscous small-scale motions near the wall. Chapman (1979) estimated that the number of grid points needed for LES to resolve these near-wall small-scale motions is approximately proportional to $Re_L^{1.8}$. Several LES studies have been applied to study ABL flow and dispersion in urban areas (Kanda et al., 2004; Santiago et al., 2010; Xie et al., 2008). In spite of the fact that LES computations are feasible and more accurate than Reynolds-Averaged Navier-Stokes simulation (see next paragraph), they are still very expensive.

Reynolds-Averaged Navier-Stokes simulation (RANS). As the name implies, the RANS approach solves the 'averaged' Navier-Stokes equations. In this approach, only the ensemble averaged flow properties are resolved with all other scales of eddies being modeled. The turbulent stresses required for the closure of the Reynolds-averaged momentum equation, known as the Reynolds stresses, represent the mean momentum fluxes induced by turbulence. The classical approach to modeling this term is to adopt the eddy viscosity concept originally proposed by Boussinesq (1877), which assumes a linear constitutive relationship between the turbulent stresses and the mean strain-rate tensors.

As additional equations, several types of turbulence models permit the obtainment of an estimate for the Reynolds stresses in the RANS equations: Mixing length model (Prandtl,

1925), $k-\varepsilon$ turbulence models (standard, renormalization group (RNG), realizable) (Launder and Spalding, 1974; Yakhot et al., 1992), $k-\omega$ turbulence models (Kato and Launder, 1993), Algebraic stress models (Baldwin and Lomax, 1978) and Reynolds stress models (Launder et al., 1975). The computational cost of RANS is independent of the Reynolds number, except for wall-bounded flows where the number of grid points required in the near-wall region is proportional to $ln(Re_L)$ (Pope, 2000). Although RANS is less accurate, because of its computational efficiency, RANS is the most commonly used CFD methodology for the simulation of turbulent flows encountered in industrial and engineering applications. Note that there is no turbulence model that is universally valid. For the purposes of this work, the turbulence is parameterized by the well known standard $k-\varepsilon$ closure.

1.1. Best Practice Guidelines

The accuracy of CFD is an important matter of concern. Care is required in the geometrical implementation of the model, in grid generation and in selecting proper solution set-up and parameters. Since a large number of choices needs to be made by the user in CFD simulations, some guidelines on industrial applications have been published in order to clarify the method for validation and verification of CFD results (e.g., ERCOFTAC (European Research Community on Fluids, Turbulence And Combustion) organizations' guidelines (Casey and Wintergerste, 2000)). In 2007, European Cooperation in Science and Technology (COST action 732 research group) (Franke et al., 2007) compiled a set of specific recommendations for the use of CFD in wind engineering from a detailed review of the literature. In 2008, a Working Group in the Architectural Institute of Japan (AIJ) (Tominaga et al., 2008), similar to COST 732, conducted extensive best practice advice for CFD prediction of the pedestrian wind environment around buildings. These documents primarily focused on steady RANS simulations. Here, I briefly present some guidelines for CFD in urban aerodynamics which are mainly based on COST (Franke et al., 2007) and AIJ (Tominaga et al., 2008) recommendations.

Error in CFD Simulations

In typical CFD simulations, different kinds of errors can have a very large impact on the results. Some sources of errors can be classified:

Physical Modeling Errors. Physical modeling errors are due to uncertainties in the formulation and to deliberate simplifications of the model: for instance, the RANS equations in combination with a given turbulence model, the eddy viscosity model or Boussinesq hypothesis, use of specific constants in the $k-\varepsilon$ model, use of wall functions, modeling of the surface roughness, simplifications of the geometry, etc. In general, physical modeling errors can be examined by performing validation studies that focus on certain phenomena (e.g., turbulent boundary layers).

Computer Round-Off Errors. Computer round-off errors develop with the representation of floating point numbers and the accuracy at which numbers are stored. With advanced computer resources, numbers are typically stored with 16, 32, or 64 bits. Computer round-off errors are not considered significant when compared with other errors. If they are suspected to be significant, one can perform a test by running the code at a higher precision. For simple flows, single precision runs (32 bit arithmetic) are usually adequate for convergence. In some more difficult cases, where there may be extremes of scales in the problem or very fine meshes, the use of double precision (64 bit arithmetic) can be required. This will require more memory, but may not add a huge overhead on computational time, depending on the nature of the hardware being used. The computer numbering format in the CFD code *Code_Saturne* used in this work is double precision.

Iteration-Convergence Error. This error is introduced because the iterative procedure to reach the steady state solution has to be stopped at a certain moment in time. The default values for convergence in most commercial codes are not strict because code vendors want to stress calculation efficiency. Therefore, stricter convergence criteria are required to check that there is no change in the solution. COST 732 (Franke et al., 2007) suggests that scaled residuals should drop by at least 4 orders of magnitude. AIJ (Tominaga et al., 2008) points that the suitable convergence values are largely dependent on flow configuration and boundary conditions, so it is better to check the solution directly using different convergence criteria. In this work, a standard residual value (10^{-9}) is kept and the convergence of the solution with monitoring points is checked.

Spatial and Temporal Discretization Errors. These errors are generated from representing the governing equations on a mesh that represents a discretized computational domain. For unsteady calculations also time discretization causes discretization errors. ERCOFTAC report (Casey and Wintergerste, 2000) indicates that the spatial and temporal discretization are probably the most crucial source of numerical errors. The COST 732 report (Franke et al., 2007) advises that grid sensitivity analysis is a minimum requirement in a CFD simulation. In Section (1.2.), the issues relative to the computational grid are discussed in more detail. To assess the influence of the timestep on the results, a systematic reduction or increase of the timestep should be made, and the simulation repeated. In advection dominated problems, the timestep Δt (s) should satisfy the following criteria:

$$\Delta t = CFL\,\Delta x_{min}/U_{max} \tag{2.1}$$

where Δx_{min} (m) is the minimum grid width, U_{max} $(m\,s^{-1})$ is the maximum velocity and CFL is the Courant-Friedrichs-Lewy number (Courant et al., 1967). Choosing the minimum grid spacing and the maximum velocity makes this estimate conservative. The generally suggested criteria is that $CFL < 1$.

Besides the above mentioned errors, due to a lack of information about physical parameters used within the model, the influence of the unwise choice of these parameters can also lead to errors in the results if the choice is inadequate.

Generally, many errors are made by CFD users because of lack of knowledge. As a result, simulation results can only be trusted or used if they have been performed on a mesh obtained by grid-sensitivity analysis, performed whilst taking into account the proper guidelines that have been published in the literature and carefully validated. Validation means systematically comparing CFD results with experiments to assess the performance of the physical modeling choices.

Choice of the Computational Domain

The size of the entire computational domain in the vertical, lateral and flow directions depends on the area that shall be represented and on the boundary conditions that will be used. For urban areas with multiple buildings, both COST 732 (Franke et al., 2007) and AIJ (Tominaga et al., 2008) reports suggest that the top boundary should be set $5H_{max}$ or above the tallest building with height H_{max} (Fig. 2.1). The reason is that the large distances given above the obstacles are necessary to prevent an artificial acceleration of the flow over the buildings.

After COST 732 (Franke et al., 2007), the lateral boundaries should be at a distance of $5H_{max}$ from the obstacles. Same distance should be set between the inlet boundary and the first building which allows for a fully developed flow (Fig. 2.1). The outflow boundary should be positioned at least $15H_{max}$ behind the last building to allow for flow re-development behind the wake region (Fig. 2.1). Similar requirements for the lateral and the inlet boundaries were suggested by AIJ (Tominaga et al., 2008). However, they report that there is a possibility of unrealistic results if the computational region is expanded without representation of surroundings, then the recommended outflow boundaries is at least $10H_{max}$.

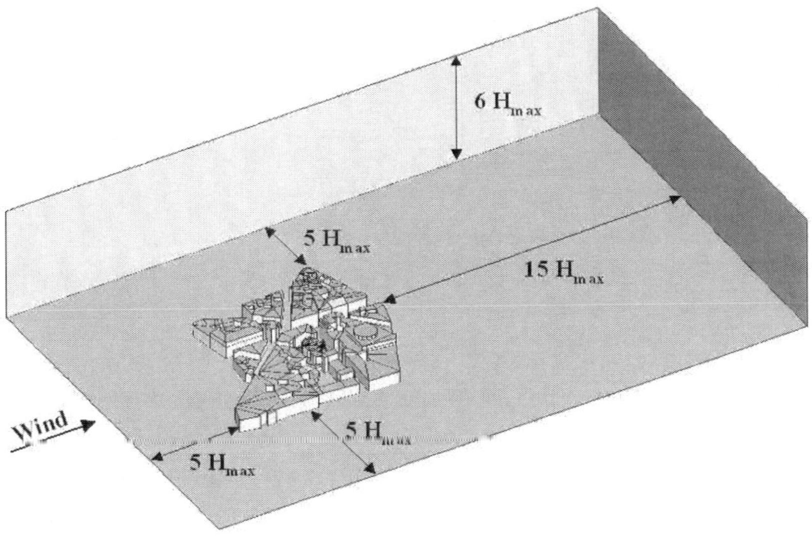

Figure 2.1. Recommended computational domain size where H_{max} refers the maximum height of the building, adapted after Franke et al. (2007) and Tominaga et al. (2008).

Initial and Boundary Conditions

Incorrect or inappropriate specification of initial or boundary conditions is a very common cause of errors. It may lead to the solution of the wrong problem as well as convergence difficulties.

Initial Conditions. Initial data and inflow data are very often chosen the same. This is a good starting point for most models. Commencing with the larger-scale field, which is expected to be close to the final solution, will reduce the computational efforts needed to reach stationary solutions. However, if these initial data are not close to the real initial conditions (e.g., wrong wind direction) then an accurate solution can not be expected. Since initial data are not known perfectly, but include uncertainties that result from lack of measurement or measurement inaccuracy, the initial input values are never perfectly known. Therefore COST 732 (Franke et al., 2007) advises keeping initial data uncertainty to as little as possible and to keep in mind that the initial data influence the model results in unsteady simulations.

Inlet Boundary Conditions. The proper choice of boundary conditions is very important. Since they represent the influence of the larger-scale surroundings and they determine to a large extent the solution inside the computational domain. At the inlet boundary, the mean velocity profile is often obtained from the academic logarithmic profile modeling the flow over the upwind terrain via the roughness length z_0 (m), or from the profiles of the wind tunnel simulations. In simulation of field experiments, available information from nearby meteorological stations is used to determine the wind speed U_{ref} $(m\ s^{-1})$ at a reference height z_{ref} (m) (Stull, 1988).

In the case the vertical distribution of turbulent energy $k(z)$ $(m^2\ s^{-2})$ is not available in the data set, COST 732 (Franke et al., 2007) assuming a constant friction velocity in ABL, suggests:

$$k(z) = \frac{U^{*2}_{ABL}}{\sqrt{C_\mu}}, \tag{2.2}$$

and the dissipation rate $\varepsilon(z)$ can be expressed as:

$$\varepsilon(z) = \frac{U^{*3}_{ABL}}{\kappa(z+z_0)}, \tag{2.3}$$

where U^{*}_{ABL} $(m\ s^{-1})$ represents the atmospheric boundary layer friction velocity, C_μ is a constant coefficient $(= 0.09)$, κ is the von Kármán constant $(= 0.4)$.

Nevertheless, AIJ (Tominaga et al., 2008) points out that, in their recommendations, COST 732 (Franke et al., 2007) assumes that the height of the computational domain is much lower than the atmospheric boundary layer height, since the assumption of a constant friction velocity is only valid in the lower part of the atmospheric boundary layer - surface boundary layer (Stull, 1988). Therefore, AIJ (Tominaga et al., 2008) proposes the following estimation equation between the vertical profile of turbulent intensity $I(z)$ and turbulent energy $k(z)$:

$$k(z) = I(z)^2 U(z)^2, \tag{2.4}$$

with

$$I(z) = 0.1\left(\frac{z}{z_G}\right)^{-\beta-0.05}, \tag{2.5}$$

where $U(z)$ is the vertical profile of velocity (ms^{-1}), z_G (m) is the boundary layer height and β is the power-law exponent. Both z_G and β are determined by terrain category (see Table 2.1).

Table 2.1. Parameters determining $I(z)$; Modified after AIJ (2004); Category I: open, no significant obstruction, sea, lake; II: open, few obstructions, grassland, agricultural field; III: suburban, wooded terrain, few tall building (4 to 9-story); IV: city, tall buildings (4 to 9-story); V: city, heavy concentration of tall buildings (higher than 10-story).

Category	I	II	III	IV	V
z_G	250	350	450	550	650
β	0.1	0.15	0.2	0.27	0.35

AIJ (Tominaga et al., 2008) also recommends that the values of the dissipation rate be given by assuming local equilibrium of production term for k equation. When the vertical gradient of velocity can be expressed by a power law with exponent β:

$$\varepsilon(z) = C_\mu^{1/2} k(z) \frac{U_{ref}}{z_{ref}} \beta \left(\frac{z}{z_{ref}}\right)^{\beta-1}. \tag{2.6}$$

Top Boundary Conditions. AIJ (Tominaga et al., 2008) reports that if the computational domain is large enough (Fig. 2.1), the boundary conditions for lateral and top boundaries do not have significant influences on the calculated results around the target buildings. However, COST 732 (Franke et al., 2007) reports stress the importance of the choice of the top boundary condition and lateral boundary conditions. If symmetry boundary conditions are applied to the top boundary, these might enforce a parallel flow, by forcing the velocity component normal to the boundary to vanish. Furthermore, prescribing zero normal derivatives for all other flow variables may lead to a change from the inflow boundary profiles (which can have a non-zero gradient at the height of the top of the domain). On the other hand, if the top boundary is handled as an outflow boundary, it can allow a normal velocity component at this boundary. In order to prevent a horizontal change from the inflow profiles, it is recommended to prescribe a constant shear stress (velocity gradient) at the top, corresponding to the inflow profiles. The latter option is taken in this work.

Lateral Boundary Conditions. In the CFD codes, when the approach flow direction is parallel to the lateral boundaries, symmetry boundary conditions are frequently used at lateral boundaries. This option is used in an idealized case simulation (see Chapter 3). COST 732 (Franke et al., 2007) states that symmetry boundary conditions enforce a parallel flow by requiring a vanishing normal velocity component at the boundary. Therefore, the boundary should be positioned far enough from the built-up area of interest in order not to lead to an artificial acceleration of the flow near the lateral boundaries (Fig. 2.1). In cases where different wind directions are to be simulated with the same computational domain, the lateral boundaries become inflow or outflow boundaries. This is the case in Chapter 3, 4 and 5.

Outlet Boundary Conditions. At the boundary behind the obstacles (where all or most of the fluid leaves the computational domain), open boundary conditions are mostly used in CFD simulations. The open boundary conditions are either outflow or constant static pressure boundary conditions. The outflow boundary conditions are applied in this work. With an outflow boundary condition, the derivatives of all flow variables are forced to zero, corresponding to a fully developed flow. COST 732 (Franke et al., 2007) indicates that this boundary should be ideally far enough from the last building in order not to have any fluid re-entering into the computational domain. This also applies when using a constant static pressure at the outflow boundary, with the derivatives of all other flow variables forced to vanish. Imposed pressure at the outlet is used in *Code_Saturne*.

Wall Boundary Conditions. At solid walls, the no-slip boundary condition is used for the velocities. COST 732 (Franke et al., 2007) mentions two different approaches to resolve the shear stress at smooth walls. The first one is the low-Reynolds number approach which resolves the viscous sublayer and computes the wall shear stress from the local velocity gradient normal to the wall. The equations for the turbulence quantities contain damping functions to reduce the influence of turbulence in this region dominated by molecular viscosity. The low-Reynolds number approach requires a very fine mesh resolution in the wall-normal direction. The first computational node should be positioned at a dimensionless wall distance z^+ given by:

$$z^+ = zu_*/\nu \approx 1, \tag{2.7}$$

where z (m) is the distance normal to the wall, ν $(m^2\ s^{-1})$ is the kinematic viscosity and u_* $(m\ s^{-1})$ is the friction velocity, computed from the time-averaged wall shear stress τ_w $(N\ m^{-2})$:

$$u_* = (\tau_w/\rho)^{1/2}, \tag{2.8}$$

with ρ $(kg\ m^{-3})$ the density.

To reduce the number of grid points in the wall-normal direction and therefore, the computational costs, another approach called wall functions is applied as an alternative approach to compute the wall shear stress. With the wall function approach, the wall shear stress is computed assuming a logarithmic velocity profile between the wall and the first computational node in the wall-normal direction. For the logarithmic profile to be valid, the first computational node should be placed at a dimensionless wall distance of z^+ between 30 and 500 for smooth walls. Also, for wall function modeling the turbulence quantities have to be modified at the first computational node. They are usually calculated assuming an equilibrium boundary layer, consistent with the logarithmic velocity profile. In spite of invalidity in regions of flow separation, of reattachment and of strong pressure gradients and also unpredictable of the transition from laminar to the turbulent boundary, the effect of wall functions on the solution away from the wall is, however, regarded as small in the built environment.

Furthermore, the wall function approach is also used for rough walls. Blocken et al. (2007) state different wall functions and demonstrate the importance of four basic requirements for CFD simulation of ABL flow with sand-grain wall functions. The four requirements are:

- a high mesh resolution in the vertical direction near the bottom of the computational domain,
- the horizontal homogeneity of ABL flow in the upstream and downstream region of the domain,
- a distance y_P (m) from the center point P of the wall-adjacent cell to the wall (bottom of the domain) that is larger than the physical roughness height k_S (m) of the terrain $(y_P > k_S)$,
- the relationship between the equivalent roughness height k_S and the corresponding aerodynamic roughness length z_0 (m).

In order to deal with the problem of the impossibility of simultaneously satisfying all four requirements in the k_s type wall functions for fully rough surfaces (i.e., standard wall functions modified for roughness based on experiments with sand-grain roughness), Blocken et al. (2007) consider that the best solution is to violate the third requirement $y_P > k_S$ and advise assessing the extent of horizontal inhomogeneity by a simulation in an empty computational domain prior to the simulation domain with obstacles. A roughness wall function is used in this work and is presented in Section 3.5.. The similar consideration to the thermal boundary layer for heated walls is also applied. In this work, the dimensionless value z^+ is between 2.8 and 268 in Chapter 3 and between 0.2 and 14 in Chapter 4.

Algorithmic Considerations

In order to be numerically solved, the basic equations have to be discretized and transformed into algebraic equations. For time-dependent problems, second-order methods should also be chosen for the approximation of the time derivatives. Higher order advection differencing schemes can lead to numerical oscillations that may cause poor convergence, or have quantities to overshoot. Running with first order upwind schemes may help to overcome this. However, it should be remembered that the spatial gradients of the transported quantities tend to become diffusive due to a large numerical viscosity of the upwind scheme. The reports of both COST 732 (Franke et al., 2007) and AIJ (Tominaga et al., 2008) do not recommend the use of first-order methods like upwind scheme except in initial iterations.

In this research, the author first performed the simulations with a center scheme. However, when the thermal effects are taken into account in a low wind speed case, using a center scheme happens to create numerical instabilities, especially in the inflow region, thus an upwind scheme is used.

1.2. Grid Issues

The discrete spatial domain (either for Finite-Difference, Finite-Volume or Finite-Element methods) is known as the grid or mesh. Mesh generation is often considered as the most important and most time-consuming part of CFD simulations. The quality of the mesh plays a direct role in the quality of the simulations, regardless of the flow solver used. Additionally, the solver will be more robust and efficient when using a well constructed mesh.

Grid Classification

As CFD has developed, better algorithms and more computational power have become available, resulting in a diversification in solver techniques. One direct result of this development has been the expansion of available mesh elements and mesh connectivity (how cells are connected to one another). The elements in a mesh can be classified in various ways. Based on the connectivity of the mesh, they can be classified: structured or unstructured. Structured grid generators are most commonly used when strict elemental alignment is mandated by the analysis code or is necessary to capture the physical phenomenon. Unstructured mesh generation, on the other hand, relaxes the node valence requirement, allowing any number of elements to meet at a single node. *Code_Saturne* can work with both a structured grid and an unstructured mesh. Another grid classification is based on the dimension and type of the elements. Common elements in 2D are triangles or rectangles, and common elements in 3D are tetrahedral or hexahedral. Here I briefly describe the types of meshes which are commonly used.

Hexahedral Meshes. Hexahedral meshes (either structured or unstructured grids) take their name from the fact that the mesh is characterized by a polyhedron with six faces. Although the element topology is fixed, the mesh can be shaped to be body fitted through stretching and twisting of the grid. Hexahedral meshes have the advantage of allowing a high degree of control. Indeed, hexahedral grids, which are very efficient at filling space, support a high degree of skewness and stretching before the solution is significantly affected. Also, the mesh can be flow-aligned, thereby yielding to greater accuracy of the solver. Hexahedral mesh flow solvers typically require lower amount of memory for a given mesh size and execute faster because they are optimized for the structured layout of the mesh. Lastly, post-processing of the results on a hexahedral block mesh is typically a much easier task. Because the logical mesh planes make excellent reference points for examining the flow field and plotting the results.

Compared to tetrahedral meshes (see next paragraph), for the same cell count, hexahedral meshes will give more accurate solutions, especially if the grid lines are aligned with the flow. The major drawback of hexahedral meshes is the time and expertise required to lay out an optimal block structure for an entire model. Some complex geometries (see the CAPITOUL mesh in Chapter 4) are very hard even impossible to mesh with hexahedral block topologies. In these areas, the user is forced to stretch or twist the elements to a degree which drastically affects solver accuracy and performance. With the present computational power, mesh generation times are usually measured in hours if not days. This type of the mesh is used in the simple building geometry case (cf. Chapter 3 one mesh with about 55 000 cells and one with about 253 000 cells).

Tetrahedral Meshes. Tetrahedral meshes (always unstructured grids) are characterized by irregular connectivity which is not readily expressed as a 3D array in computer memory, but use an arbitrary collection of elements to fill the domain. Tetrahedral meshes can be stretched and twisted to fit the domain. These methods have the ability to be automated to a large degree. Given a good Computer-Aided Design (CAD) model, a good mesher can automatically place triangles on the surfaces and tetrahedral in the volume with very little input from the user. The advantage of tetrahedral mesh methods is that they are very automated and, therefore, require little user time or effort. And the user does not need to worry about laying out block structure or connections. Mesh generation times are usually measured in minutes or hours.

The major drawback of tetrahedral meshes is the lack of user control when laying out the mesh. Typically any user involvement is limited to the boundaries of the mesh with the mesher automatically filling the interior. Triangle and tetrahedral elements have the problem that they do not stretch or twist well, therefore, the mesh is limited to being largely isotropic, i.e., all the elements have roughly the same size and shape. This is a major problem when trying to refine the mesh in a local area, often the entire mesh must be made much finer in order to get the point densities required locally. Another drawback of the methods is their reliance on good CAD data. Most meshing failures are due to some (possibly minuscule) error in the CAD model. Tetrahedral flow solvers typically require more memory and have longer execution times than structured hexahedral mesh solvers on a similar geometry. Post-processing the solution on a tetrahedral mesh requires powerful tools for interpolating the results onto planes and surfaces of rotation for easier viewing. Since *Code_Saturne* accepts meshes with any type of cell and any type of grid structure and a CAD data is available, this type of the mesh in CAPITOUL studies is used (cf. the CAPITOUL mesh in Chapter 4 with about 4.5 million cells).

Hybrid Meshes. A hybrid mesh is a mesh that contains hexahedral portions and tetrahedral portions. Hybrid meshes are designed to take advantage of the positive aspects of both hexahedral and tetrahedral meshes. They use some forms of hexahedral cells in local regions while using tetrahedral cells in the bulk of the domain. Hybrid meshes contain hexahedral, tetrahedral, prismatic, and pyramid elements in 3D and triangles and quadrilaterals in 2D. The various elements are used according to their strengths and weaknesses. Hexahedral elements are excellent near solid boundaries (where the gradients are high) and afford the user a high degree of control, but are time-consuming to generate. Prismatic elements (usually triangles extruded into wedges) are useful for resolving near wall gradients, but suffer from the fact that they are difficult to cluster in the lateral direction due to the underlying triangular structure. In almost all cases, tetrahedral elements are used to fill the remaining volume. Pyramid elements are used to transition from hexahedral elements to tetrahedral elements. Many codes try to automate the generation of prismatic meshes by allowing the user to define the surface mesh and then marching off the surface to create the 3D elements. While very useful and effective for smooth shapes, the extrusion process can break down near regions of high curvature or sharp discontinuities. The advantage of hybrid mesh methods is to control the shape and distribution of the grid locally, which can yield excellent meshes. The disadvantage is that they can be difficult to use and require user expertise in laying out the various grid locations and properties to get the best results. The generation of the hexahedral portions of the mesh will often fail due to complex geometry or user input errors. While the flow solver will use more resources than a structured hexahedral block code, it should be very similar to an unstructured tetrahedral code. Post-processing the flow field solution on a hybrid grid suffers from the same disadvantages as a tetrahedral mesh. The time required for mesh generation is usually measured in hours or days. This kind of the mesh is used in EUREQUA case (cf. the EUREQUA mesh in

Chapter 4 with about 6 million cells).

Choice of the Computational Mesh

With the Finite Volume, Finite Difference and Finite element methods the computational results depend crucially on the mesh that is used to discretize the computational domain. A high quality mesh should facilitate the capturing of the important physical phenomena like shear layers or vortices with sufficient resolution and no large errors introduced.

Geometrical Representation of Obstacles. The level of detail required for individual buildings or obstacles depends on their distance from the central area of interest. COST 732 (Franke et al., 2007) points out that the central area of interest should be reproduced with as much detail as possible. This naturally increases the number of cells that are necessary to resolve the details. The available computational resources, therefore, limit the details which can be reproduced. Nevertheless, the numerical studies do not always require a very high degree of detail. In this work, buildings are represented as simple blocks (cf. Chapter 3, the MUST experiment or with more details in Chapter 4 the CAPITOUL experiment).

Mesh Resolution. When a global systematic mesh refinement is not possible due to resource limitations, at least a local mesh refinement should be used in the areas of interest. Grid stretching/compression should be small in regions of high gradients to keep the truncation error small. In these regions, both COST 732 (Franke et al., 2007) and AIJ (Tominaga et al., 2008) advise an expansion ratio of 1.3 or less. AIJ (Tominaga et al., 2008) suggests that the minimum grid resolution should be set to about $1/10$ of the building height scale (about 0.5 to 5.0 m) within the region including the evaluation points around the target building. Moreover, the evaluation height (1.5 to 5.0 m above ground) should be located at the third or higher grid cell from the ground surface. COST 732 (Franke et al., 2007) suggests that at least 10 cells should be used per building side and 10 cells per cube root of building volume as an initial choice. It is also recommended that pedestrian wind speeds at 1.5 to 2 m height should be calculated at the third or fourth cell above the ground. However, using only 5 cells per building side, simulation results presented in Chapter 3 are still acceptable. For the results presented in Chapter 4 and 5, the simulations run on computer clusters, so at least 10 cells for the target building are used.

The mesh should be generated with consideration of such things as resolution, density, aspect ratio, stretching, orthogonality, grid singularities, and zonal boundary interfaces. However, the sensitivity of the results on the mesh resolution should be tested. COST 732 (Franke et al., 2007) and AIJ (Tominaga et al., 2008) indicate that the number of fine meshes should be at least 1.5 times the number of coarse meshes in each dimension, and at least three refined meshes should be tested. Additionally, for the unstructured mesh, it is necessary to ensure that the aspect ratios do not become excessive in regions adjacent to coarse grids or near the surfaces of complex geometries. For improved accuracy, it is recommended to arrange the boundary layer elements (prismatic cells) parallel to the walls or the ground surfaces (Fig. 2). Both COST 732 (Franke et al., 2007) and AIJ (Tominaga et al., 2008) introduce the same technique.

2. Review of Some Urban Surface Energy Balance Models

2.1. Urban Surface Energy Balance Modeling

The behavior of the atmospheric Urban Canopy Layer (UCL) is the result of the interactions between atmospheric structures induced by the urban heterogeneities. One important feature of the UCL is the urban surface energy balance. In recent years, with increasing computer power and development of micrometeorological parameterizations, Surface Energy Balance (SEB) models have evolved rapidly and increased in complexity.

A large number of models now exist with different assumptions about the important features of the surface and exchange processes that need to be incorporated. They can be classified into five categories, depending on the complexity of the parametrization, each one having its advantages and weaknesses (Masson, 2006):

- Empirical models: this type of approach makes it possible to use extremely simple schemes. For instance, the Local-scale Urban Meteorological Parameterization Scheme (LUMPS) (Grimmond and Oke, 2002) is a local-scale urban meteorological parameterization scheme capable of predicting the 1D spatial and temporal variability in heat fluxes in urban areas. Their main weakness is that they are based on statistics from field data, therefore they are limited to the range of conditions (land cover, climate, season, etc.) encountered in the original studies.

- Vegetation models without drag terms: this type of approach is based on the observation that roughness lengths and displacement heights are large over cities. Some refinement, depending on how the buildings are spatially organized, can be used to evaluate the roughness lengths. When coupled to an atmospheric model, the first atmospheric level is above the surface scheme, with all the friction located at this level. Grimmond and Oke (1999a) analyze nature, sensitivity, and size of aerodynamic parameters obtained using morphometric methods, especially in the context of the physical structure of parts of North American cities.

- Vegetation models with drag terms: these models are derived from forest canopy parameterizations. A drag force is directly added in the equations of motions in the atmospheric model, up to the height of the highest buildings. For instance, when reviewing the forest canopy parameterizations of Yamada (1982), Brown and Williams (1998) replace the fraction of the grid cell covered by trees with urban structures and use drag coefficients to incorporate the urban canopy parameterizations into a mesoscale code. Additional terms in the turbulence equation can also be taken into account. The main disadvantage of drag based schemes is that they imply direct modification of the equations of the atmospheric models to which they are coupled. A detailed urban and rural canopy parameterization developed inside the Penn State/NCAR Mesoscale Model (MM5) (Grell et al., 1994), DA-SM2-U (Dupont et al., 2004) uses the drag-force approach representing the dynamic and turbulent effects of the buildings and vegetation, and a modified version of the single layer scheme (see next paragraph), the Soil Model for Submesoscales, Urbanized Version (SM2-U) (Dupont et al., 2002; Dupont and Mestayer, 2006), to represent the thermodynamic effects of the canopy elements.

- Single layer schemes: in this approach, the exchanges between the surface and the atmosphere occur at the top of the canopy. This means that when this scheme is coupled with an atmospheric model, the first level of the atmospheric model is located above the roof level. This has the advantage of simplicity and transferability. In this approach, the characteristics of the air in the canopy must be parametrized. Generally, the logarithmic law for wind is assumed above the top of the canopy, and an exponential law below. Air temperature and humidity are assumed to be uniform in the canyon. One of these models is SM2-U (Dupont et al., 2002; Dupont and Mestayer, 2006), which includes a one-layer urban-and-vegetation canopy model to integrate the physical processes inside the urban canopy, such as heat exchanges, heat storage, radiation trapping, water interception, or surface water runoff. They are integrated in a simple way (e.g., neither separated walls and roads energy budgets nor wind speed parameterization inside the canopy). Another example is the Town Energy Balance (TEB) scheme of Masson (2000). TEB consists of a facet-averaged scheme with one generic roof, one generic wall, and one generic road. The advantage of the integrated resolution is that few individual SEBs need to be resolved and therefore computation time is kept low, with a simple approach to model the inner canopy wind flow. TEB has been shown to accurately reproduce the SEB from regional to mesoscale and urban scales (Masson et al., 2002; Lemonsu et al., 2004).

- Multi-layer models: in these schemes, the wind and temperature are not uniform in the canopy, they depend on the interaction between the urban surfaces and the air at different levels in the roughness sub-layer. However, such a refinement is made at the cost of direct interaction with the atmospheric models because their equations are modified. Among these models, the Building Effect Parameterization model (Martilli et al., 2002) presents a high level of detail of the SEB, since any number of road and wall orientations are available, different building heights can be taken into account, and at each level of the wall intersecting an air level, there is a separate energy budget. This feature means this model is able to represent the differential heating of the wall when the sun is close to the horizon.

In addition, 3D models generally based on view factor calculation, compute for each urban sub-facet the incoming radiative fluxes and simple convection parameterization or coupled with CFD. The SOLENE (Miguet and Groleau, 2002) and 3D-Computer Aided Design (Asawa et al., 2008) models are based on a realistic description of the canopy structure using a geometrical 3D surface model assigning radiative and thermal properties to each subfacet of the model and a constant transfer coefficient for each class of elements. The model is originally designed for simulating sunshade, natural lighting, and heat transfer for architectural purposes. The Discrete Anisotropic Radiative Transfer (DART) model (Gastellu-Etchegorry et al., 2004) simulates the radiative transfer in the whole optical domain simultaneously in the atmosphere and in the urban and vegetated landscapes, with or without topography. A major feature of DART is that it can simulate images in the plane of the sensor, for different altitudes from the bottom to the top of the atmosphere. The new version of the DART model, DART-EB (Gastellu-Etchegorry, 2008), includes an energy balance component. In the case of urban canopies, turbulent fluxes and conduction are computed with classical boundary layer laws, using the equations of the TEB model (Masson, 2000). The Simple Urban Energy Balance Model for Mesoscale Simulations (SUMM) (Kanda et al., 2005), which represents the urban canopy with an infinitely extended regular array of uniform buildings, is more adapted for the mesoscale. The Temperatures of Urban Facets in 3D (TUF-3D) model (Krayenhoff and Voogt, 2007) uses the radiosity approach based on interpatch view factors to model radiative exchange between the identical square patches that compose the simplified 3D urban geometry. An exponential inner canopy wind speed profile is employed. TUF-3D has applications in both surface temperature distributions and thermal remote sensing anisotropy at several scales.

In view of a wide range of urban energy balance models, it is not possible to single out one universal SEB model, which would be valid for all cases. However, a classification and a comparison of these models can be very helpful to identify the models for understanding the complexity required to model energy and water exchanges in urban areas. To do so, Grimmond et al. (2010, 2011) conducted an international Urban Energy Balance Models Comparison which is presented in the next section.

2.2. International Urban Energy Balance Models Comparison

Grimmond et al. (2010, 2011) conducted an international comparison of more than 30 models based on modeling the methodology and initial results with the same experimental data set. The fundamental requirement for the models to be included in this international comparison study is that they simulate urban energy balance fluxes. These schemes have varying levels of complexity, and model different fluxes. Table 2.2 refers to all the models (some models have different versions) involved in the international comparison study. While all the models calculate the outgoing radiative fluxes $S^{\uparrow}$ and $L^{\uparrow}$, net all-wave radiation Q^* and turbulent sensible heat flux Q_H, some do not model either turbulent latent heat flux Q_E or the additional sources of energy Q_F, and some model neither. It is noteworthy that some models solve the heat conduction equation using the force-restore method, while others solve the one-dimensional heat conduction equation. Both of these two approaches have been implemented into the developed model (cf. section 3.4.).

The evaluation of these models shows that, overall, they model accurately Q^*, while they are less capable of modeling Q_E. There is evidence that some classes of models perform better for individual fluxes but not overall. Typically, those that perform best during daytime do not perform best at night. For most models, the mean bias error (MBE) for sensible heat flux is positive, which might be explained by observational errors.

Some of the greatest differences in model performance are found between classes of model that treat vegetation and reflections differently. Some of the smallest differences relate to approaches used to calculate heat storage flux and urban morphology. Not including vegetation, even for a site with limited vegetation, results in the poorest performance for all fluxes during the day (in terms of the root mean square error (RMSE)) and accounts for latent heat flux at night. In general, using a bulk albedo/emissivity results in better performance for all fluxes during the day.

Table 2.2. Urban energy balance models with a given reference utilizing each model participating in the International Urban Balance Models Comparison Project. Modified after Grimmond et al. (2010)

Code	Model Name	References (e.g.,)
BEP02	Building Effect Parameterization	Martilli et al. (2002)
BEP_BEM08	BEP coupled with Building Energy Model	Salamanca et al. (2009)
CLMU	Community Land Model - Urban	Oleson et al. (2008)
GCTTC	Green Cluster Thermal Time Constant model	Shashua-Bar and Hoffman (2004)
IISUCM	Institute of Industrial Science Urban Canopy Model	Kawamoto and Ooka (2009)
JULES	Joint UK Land Environment Simulator	Best et al. (2006)
LUMPS	Local-scale Urban Meteorological Parameterization Scheme	Offerle et al. (2003)
NKUA	University of Athens Model	Dandou et al. (2005)
MORUSES	Met Office Reading Urban Surface Exchange Scheme	Harman and Belcher (2004)
MUCM	Multi-layer Urban Canopy Model	Kondo et al. (2005)
NJU-UCM-S	Nanjing University Urban Canopy Model-single layer	Kusaka et al. (2001)
NJUC-UM-M	Nanjing University Urban Canopy Model-multiple layer	Kanda et al. (2005)
NSLUCM /NSLUCMK /NSLUCM-WRF	Noah land surface model/Single-layer Urban Canopy Model	Chen et al. (2009)
SM2U	Soil Model for Submesoscales (Urbanized)	Dupont and Mestayer (2006)
SNUUCM	Seoul National University Urban Canopy Model	Ryu et al. (2009)
SRUM2/SRUM4	Single Column Reading Urban Model tile version	Harman and Belcher (2006)
SUEB	Slab Urban Energy Balance Model	Fortuniak et al. (2005)
SUMM	Simple Urban Energy Balance Model for Mesoscale Simulation	Kawai et al. (2007)
TEB	Town Energy Balance	Masson (2000)
TEB07	Town Energy Balance	Hamdi and Masson (2008)
TUF2D	Temperatures of Urban Facets 2D	Krayenhoff and Voogt (2007)
TUF3D	Temperatures of Urban Facets 3D	Krayenhoff and Voogt (2007)
VUCM	Vegetated Urban Canopy Model	Lee and Park (2004)

Classifications show that no model includes all the characteristics. In general, the simpler models perform as well as the more complex ones, based on all statistical measures, and no individual model performs best for all the fluxes. This finding has very significant implications for the applications of any model. It may also imply that, in some cases, models perform well but for the wrong physical reasons. For example, if a model overestimates the net short-wave radiation, but accurately models the sensible heat flux, it may indicate a problem in the physical representation of the heat exchanges between the surfaces and the atmosphere (since it needs to absorb more energy to get the right sensible heat flux).

When considering each individual characteristic, the models that best perform are those that are the simplest, since they need to assign only one parameter, which can be chosen close to the observed value. Considering all the characteristics together, the simplest and the most complex models have similar results and perform better than the medium complexity models. Additional surface information is important for improving model performance. Simpler models often show a net improvement with additional information; the more complex models do not. This may be because there was not enough additional detailed information provided, so it was more difficult for the users to decide how to properly use this information. It is expected that more complex models may have more potential for future improvements, as they are able to resolve more details without deteriorating their performance. The most complex models are more flexible and have the potential to describe the physical interactions between the atmosphere and the urban surfaces.

3. A New Coupled Thermal-Radiative-Dynamic 3D Scheme in *Code_Saturne* for Modeling Urban Areas

3.1. Choice of the Model

Recent researches have sought to reconsider the problem of modeling the SEB, particularly to improve the modeling of the thermal-radiative and aerodynamic phenomena. For instance, in order to learn more about the impacts of different proportion of green area, Jesionek and Bruse (2003) developed a classification scheme for typical European urban building types and have simulated them systematically with the microscale climate model ENVI-met. The simulation results where analyzed primarily in view of thermal advantages and disadvantages of increased green area and its effects on pollution dispersion and accumulation. It is shown, that especially in densely built-up block structures, green area with trees leads to higher pollution concentration, while in more open structures the thermal advantages of green area, due to shadow effects, can be fully used to improve of the microclimate. Chen et al. (2009) performed coupled simulations of convection, radiation and conduction to evaluate the outdoor thermal environment over different urban blocks, a high-rise area and a mid-rise area in the city of Tokyo in Japan, to compare the effects of measures such as the position of the heat release point of air-conditioning, greening, high surface albedo, and traffic volume. The results showed that the effectiveness of moderation countermeasures differed according to the configuration of the urban blocks. Bouyer (2009) presented the SOLENE-microclimate model including a soil model and a inner building thermal model, this last one allowing to compute the energy consumption of a building interacting with its urban environment. They are both integrated into the SOLENE, then using the commercial CFD tool (Ansys Fluent) as a quasi-dynamic coupling to calculate the air temperature, mass concentrations of moisture and heat transfer coefficient. However, only the energy transport and moisture equations in the CFD tool are solved. Except for the initialization phase, the absence of resolution of momentum equations does not allow to take into account the movement of natural and mixed convection of the airflow in the canopy.

The present research aims to accurately simulate the atmosphere and surfaces in urban environments at microscale. The objective is to fully model the 3D airflow in the urban canopy in non-neutral conditions and, therefore, to take into account atmospheric radiation and heat transfers for complex geometries. A new 3D microscale radiative scheme has been previously implemented in the open-source CFD code *Code_Saturne*, described in details by Milliez (2006). As a full thermal-radiative-dynamic coupling, the model was evaluated with idealized cases, using as a first step, a constant 3D wind field (Milliez, 2006). In this section, I describe the main features of the model and some current improvements.

3.2. Presentation of the Atmospheric Module in *Code_Saturne*

Generality of *Code_Saturne*

Developed by Électricité De France for laminar or turbulent flow, *Code_Saturne* is a CFD code which can handle complex geometry and complex physics (Archambeau et al., 2003). *Code_Saturne* solves the Navier-Stokes equations for 2D, 2D axisymmetric, or 3D, steady or unsteady, laminar or turbulent, incompressible or dilatable flows, with or without heat transfer, and with possible scalar fluctuations. Beside the atmospheric module, the code also includes a Lagrangian module, a semi-transparent radiation module, a gas combustion module, a coal combustion module, an electric module (Joule effect and electric arc) and a compressible module.

The code uses a Finite Volume discretization. A wide range of unstructured meshes, either hybrid (containing elements of different types) and/or non-conform, can be used. Within the framework of the Finite Volume approach, the numerical solver integrates the equations over each cell of the unstructured mesh. For RANS simulations, the time scheme is an implicit first order Euler approach. A fractional step scheme is used to solve the mass and momentum equations. The first step (predictor step) provides predicted velocity components: they are determined sequentially and without coupling between each other. The mass equation is taken into account during the second step (corrector step): a pressure Poisson equation is solved and the mass fluxes at the cell faces are updated.

Furthermore, the equations for the turbulent variables (turbulent kinetic energy and dissipation or Reynolds stresses and dissipation) are solved, using also the Euler approach. For the $k-\varepsilon$ model, an additional step is carried out to couple the source terms. Next, the equations for the scalars are solved, also with the Euler approach. Finally, all the variables are updated and another time step may start.

Specificity of the Atmospheric Module

The atmospheric module is adapted for multi-scale atmospheric airflow (either neutral or stratified) and pollutant dispersion studies. Detailed described in Milliez and Carissimo (2007), the atmospheric module of *Code_Saturne* solves the Navier-Stokes equations for momentum and equations for additional scalars with an anelastic approximation (i.e., by filtering the acoustic waves while keeping an assumption of compressibility, by neglecting the variations in time of the density in the continuous equation). It can take into account larger scale meteorological conditions, by using a meteorological file that contains the wind velocity, turbulence kinetic energy, dissipation rate, and temperature profiles. As currently done in atmospheric models, the potential temperature is the thermodynamic variable in the energy conservation equation, which determines stratification effects on vertical turbulent transport and allows estimation of the surface-air thermal gradient that controls convective heat transfer. An ideal gas state equation is used to take into account the water vapor content.

Two turbulent approaches are available in the module, RANS and LES. As mentioned before, the RANS approach with a $k-\varepsilon$ turbulence closure is chosen for this work. It should be stressed that despite the fact that the $k-\varepsilon$ closure is generally unable to precisely capture the geometry dependent large eddies in many complex flows and overestimates the dissipated energy, it gives a fairly acceptable accuracy with a reasonable computational time for this research work.

The model can take into account the Coriolis effects, but this study is at local scale therefore the Coriolis effects are neglected (i.e., the simulations are under large Rossby number conditions). In order to take into account the variation of density in the fluid, the Favre average is used in the model equations. For a variable v, the Favre average for RANS $\widetilde{v}$ and its fluctuation v' are defined as:

$$\widetilde{v} = \frac{\overline{\rho v}}{\overline{\rho}}, \; v' = v - \widetilde{v},$$

where $\overline{v}$ is the ensemble average.

The equations for mass, momentum and energy conservation that are solved in the atmospheric module are:

$$\frac{\partial \overline{\rho}\widetilde{u}_i}{\partial x_i} = 0, \tag{2.9}$$

$$\overline{\rho}\left(\frac{\partial \widetilde{u}_i}{\partial t} + \widetilde{u}_j \frac{\partial \widetilde{u}_i}{\partial x_j}\right) = -\frac{\partial \widetilde{p}}{\partial x_i} + \frac{\partial}{\partial x_j}\left[\mu\left(\frac{\partial \widetilde{u}_i}{\partial x_j} + \frac{\partial \widetilde{u}_j}{\partial x_i}\right) - \frac{2}{3}\mu\delta_{ij}\frac{\partial \widetilde{u}_k}{\partial x_k}\right] - \frac{\partial(\overline{\rho}\widetilde{u'_i u'_j})}{\partial x_j} + (\overline{\rho} - \rho_{ref})g_i, \tag{2.10}$$

$$\overline{\rho}\left(\frac{\partial \widetilde{\theta}}{\partial t} + \widetilde{u}_j \frac{\partial \widetilde{\theta}}{\partial x_j}\right) = \frac{\partial}{\partial x_j}\left(\frac{\lambda}{C_p}\frac{\partial \widetilde{\theta}}{\partial x_j}\right) - \frac{\partial(\widetilde{\overline{\rho}\theta' u'_j})}{\partial x_j} + S_{rad}, \tag{2.11}$$

where u_i ($m\,s^{-1}$) is the speed for coordinate i, $\widetilde{p}$ (Pa) is the average pressure, μ ($kg\,m^{-1}s^{-1}$) is the viscosity of the fluid, g_i ($m\,s^{-2}$) the gravity for coordinate i, ρ ($kg\,m^{-3}$) the fluid density and ρ_{ref} ($kg\,m\,s^{-3}$) is the time dependent hydro-static reference state, λ ($W\,m^{-1}K^{-1}$) the thermal conductivity, θ (K) the potential temperature, C_p the specific heat capacity ($J\,kg^{-1}\,K^{-1}$) and S_{rad} ($J\,K\,m^{-5}s^{-1}$) an extra thermal source.

To close equations (2.9:2.11), the Favre-Reynolds stress and the heat flux are expressed according to Boussinesq's turbulent diffusivity model:

$$-\overline{\rho}\,\widetilde{u_i'u_j'} = \mu_t\left(\frac{\partial \widetilde{u_i}}{\partial x_j} + \frac{\partial \widetilde{u_j}}{\partial x_i}\right) - \frac{2}{3}\mu_t\delta_{ij}\frac{\partial \widetilde{u_k}}{\partial x_k} - \overline{\rho}\,\frac{2}{3}\delta_{ij}k, \quad (2.12)$$

$$-\overline{\rho}\,\widetilde{\theta' u_j'} = \frac{\mu_t}{Pr_t}\left(\frac{\partial \widetilde{\theta}}{\partial x_j}\right), \quad (2.13)$$

with Pr_t the turbulent Prandtl number.

In the $k-\varepsilon$ model, μ_t is linked to the turbulent kinetic energy per unit mass k and the dissipation ε through:

$$\mu_t = C_\mu \overline{\rho}\frac{\widetilde{k}^2}{\varepsilon}, \quad (2.14)$$

with $C_\mu = 0.09$ as in Launder and Spalding (1974).

$$\overline{\rho}\left(\frac{\partial \widetilde{k}}{\partial t} + \widetilde{u_j}\frac{\partial \widetilde{k}}{\partial x_j}\right) = \frac{\partial}{\partial x_j}\left(\left(\mu + \frac{\mu_t}{\sigma_k}\right)\frac{\partial \widetilde{k}}{\partial x_j}\right) \quad (2.15)$$

$$+\widetilde{P} + \widetilde{G} - \overline{\rho}\varepsilon + S_k \quad (2.16)$$

$$\overline{\rho}\left(\frac{\partial \varepsilon}{\partial t} + \widetilde{u_j}\frac{\partial \varepsilon}{\partial x_j}\right) = \frac{\partial}{\partial x_j}\left(\left(\mu + \frac{\mu_t}{\sigma_\varepsilon}\right)\frac{\partial \varepsilon}{\partial x_j}\right)$$

$$+C_{\varepsilon 1}\frac{\varepsilon}{k}(\widetilde{P} + C_{\varepsilon 3}\widetilde{G}) - C_{\varepsilon 2}\,\overline{\rho}\frac{\varepsilon^2}{\widetilde{k}} + S_\varepsilon. \quad (2.17)$$

The equation for the turbulence and the dissipation are transport equations:
S_k and S_ε are extra source terms of turbulence and dissipation respectively that can be used for example in drag-porosity models (cf. Chapter 4 EUREQUA case) and $\widetilde{P}$ and $\widetilde{G}$ are respectively the production rate of k and the production or destruction rate due to buoyancy:

$$\widetilde{P} = -\overline{\rho}\,\widetilde{u_i'u_j'}\frac{\partial \widetilde{u_i}}{\partial x_j}, \quad (2.18)$$

$$\widetilde{G} = \frac{\mu_t}{Pr_t}\frac{1}{\overline{\overline{\theta}}}\frac{\partial \overline{\theta}}{\partial x_j}g_j. \quad (2.19)$$

The parameters of the equation are: $\sigma_k = 1$, $\sigma_\varepsilon = 1.3$, $C_{\varepsilon 1} = 1.44$, $C_{\varepsilon 2} = 1.92$ and $C_{\varepsilon 3} = 0$ for a stable stratified atmosphere ($\widetilde{G} < 0$) and $C_{\varepsilon 3} = 1$ for an unstable stratified atmosphere ($\widetilde{G} > 0$).

3.3. 3D Atmospheric Radiative Model

In SEB models, different schemes have been developed to estimate the radiative balance (shading and trapping effect of the buildings). In a transparent media, numerous approaches employ the radiosity method (Krayenhoff and Voogt, 2007) to compute the radiative flux exchanges between any kinds of two surfaces by calculating the view factors or shape factors (sky-wall, wall-wall, sky-ground, wall-ground, ground-ground). Few approaches (Milliez, 2006; Gastellu-Etchegorry, 2008) estimate the radiative fluxes by solving the Radiation Transfer Equation (RTE). The two approaches are compared in detail in Chapter 3.

A radiative atmospheric model for flat homogeneous terrain is available in *Code_Saturne*. The principle of this model is described in detail in Stephens (1984), Musson-Genon (1993, 1994) and Milliez (2006). It solves the RTE for the short- and long-wave radiation on 1D vertical arrays. For applications in built-up areas, the basic idea for developing a new microscale 3D radiative atmospheric model was to meet the following requirements: explicitly resolve the buildings in 3D, accurately reproduce the complex urban morphology and easily couple the radiative scheme with the dynamic model (Milliez, 2006). Thus, adapting to the atmosphere a radiative heat transfer scheme available for complex geometry in *Code_Saturne*, a new atmospheric 3D radiative scheme was developed in *Code_Saturne* for the urban canopy.

Radiative Transfer Equation

The thermal radiation is an electromagnetic phenomenon for heat transfer. In a physical space discretized by the Finite Volume method, this means that each volume is not only in interaction with its direct neighbors but with all visible elements. An analytical solution of the equation describing the process of radiative transfer is not possible. However, numerical methods describing the radiative heat transfer have been developed fulfilling the main aim of computing the radiative source-term in the energy conservation equation. Assuming a gray non-diffusive semi-transparent media, the RTE can be written as follows (Douce and Méchitoua, 2003):

$$div(I(\underline{x},\underline{D})\underline{D}) = -K_a(\underline{x})L(\underline{x},\underline{D}) + K_a(\underline{x})\sigma T_a^4(\underline{x})/\pi, \tag{2.20}$$

where $I(\underline{x},\underline{D})$ refers to the intensity of radiation at the location $(\underline{x})$ along the propagation direction with the vector $\underline{D}$, $K_a(\underline{x})$ is the absorption coefficient, $L(\underline{x},\underline{D})$ the monochromatic luminance ($W\ m^{-3}\ sr^{-1}$), σ the Stefan-Boltzmann constant ($5.66703 \times 10^{-8}\ W\ m^{-2}\ K^{-4}$) and T_a the air temperature (K). In a semi-transparent media, I ($W\ m^{-3}\ sr^{-1}$) and K_a can be considered independent of the wave length and are integrated over the spectrum.

The radiative heating rate q_{rad} ($W\ m^{-3}$) is given by:

$$q_{rad}(\underline{x}) = -div \int_0^{4\pi} I(\underline{x},\underline{D})\underline{D}d\Omega, \tag{2.21}$$

where $d\Omega$ represents the element of the solid angle (sr) around the direction $\underline{D}$.

Boundary Conditions. For a solid surface, the outgoing long-wave radiative flux $L^{\uparrow}$ ($W\ m^{-2}$) at a location $\underline{x}$ is composed of a reflected and emitted part under the assumption of a gray, diffusive radiation of the surface. The boundary condition for a surface is given by:

$$L^{\uparrow}(\underline{x}) = \frac{\varepsilon\sigma T_{sfc}^4}{\pi} + \frac{1-\varepsilon}{\pi}\int_{\Omega=0}^{2\pi} I(\underline{x},\underline{D})\mid \underline{D}.\underline{n}\mid d\Omega, \tag{2.22}$$

with ε the emissivity of the surface, T_{sfc} (K) the surface temperature.

Numerical Methods. The equation governing radiative transfer is of an integro-differential type. Some methods such as the zone method and the Monte Carlo method, may be excessively computing-time demanding and therefore currently not suitable for simulation of fluid flow, chemical reactions and heat transfer in large simulation domains. Approximate and rapid solution methods combining accuracy and computational efficiency are preferred especially for 3D simulations. To resolve the RTE, there are in general two different kinds of approximate numerical approaches: ray-tracing and differential methods. The essential difference in these two groups is the direction in which the transport equations are formulated.

- Ray-tracing method: 1D equations along a multitude of individual rays through the computational domain are solved. The formulation of the equations is made along straight, arbitrarily oriented rays, which are not inevitably aligned with the coordinate system of the fluid flow. Determining all traces of the rays and cells located along its way requires either a lot of computational time or a lot of memories. In addition, the vectorization and parallelization of the algorithms of ray-tracing radiation models is limited due to the different lengths of the individual rays (Knaus et al., 1997).

- Differential Methods: These are tools to transform the equation of radiative transfer (Eq. 2.20) into a set of partial differential equations which can be formulated in the Cartesian coordinate system of the fluid flow. After discretizing the equation with the Finite-Volume method, a fast iterative sparse matrix solver can be used to compute the solution, and the code for the differential radiation models can effectively be vectorized without significant increase of memory. One of the differential methods is the Discrete Ordinates Method (DOM) (Fiveland, 1984; Liu et al., 2000; Truelove, 1987) which is used in the *Code_Saturne*.

Discrete Ordinates Method (DOM)

In the Discrete Ordinates Method, the radiative transport equation is solved for a set of n discrete directions (e.g., 32 or 128 in the *Code_Saturne*). Each direction is associated with a solid angle in which the intensity is assumed to be constant. All solid angles are non-overlapping and spanning the total angle range of 4π. The integrals over the direction are replaced by numerical quadrature summed over each ordinate. Thus, the RTE (Eq. 2.20) can be transformed into a set of partial differential equations which can be approximated by a set of differential equations. The quadrature weights are determined by an approximation, in which the number of ordinates is calculated by $n^2 + 2n$. Each direction is fixed by its direction cosines. The source term of the energy conservation equation can be computed from the difference of the intensities in all directions multiplied by the weights at the cell centers and the emissivity.

In the microscale, it can be assumed that the atmosphere is not diffusive. In fact, the angular discretization and the small number of directions make the DOM less accurate than view factor models. To illustrate this, in Chapter 3, the author will discuss the advantages and weaknesses of using DOM in the developed radiative model through a comparison with SOLENE model (Miguet and Groleau, 2002) which uses the calculation of view factors or form factors. The model is adapted to the atmospheric radiation. Both short- and long-wave radiation are taken into account separately. For the short-wave radiation, direct and diffuse solar fluxes are distinguished. It should be noted that the new radiative scheme is, initially, designed to be applied in the urban canopy. Consideration of the upper part of the atmosphere is then taken into account by the boundary conditions at the top of our domain. The next two sections present the simple models for upper boundary conditions when data are not available and when not coupled with the 1D radiative model of *Code_Saturne*.

Global Solar Radiation Model

When solar rays pass through the atmosphere, there are five main types of radiation damping processes: Rayleigh scattering, diffusion by aerosols, absorption by ozone, water vapor absorption/diffusion and uniformly-mixed gas absorption. In order to evaluate the incoming global solar radiative flux, the used model is based on the Bird Clear Sky model (Bird and Hulstrom, 1981) and METSTAT model (Maxwell, 1998). The latter one is used to provide the solar radiation data for the United States.

In clear sky conditions, the direct normal solar flux S_{Dn} $(W\ m^{-2})$ is obtained as:

$$S_{Dn} = CS_0\tau_r\tau_a\tau_{oz}\tau_{aw}\tau_g, \tag{2.23}$$

where coefficient $C = 0.9751$ from METSTAT, $S_0 = 1370(W\ m^{-2})$ is the solar constant, τ_r the transmittance due to Rayleigh diffusion, τ_a the transmittance due to aerosol absorption and scattering, τ_{oz} the transmittance due to ozone absorption, τ_{aw} the transmittance due to water vapor absorption and τ_g the transmittance due to mixed gas absorption.

The solar diffusion flux by the atmosphere is considered to be isotropic and can be expressed with three terms: Rayleigh diffusion S_{dr} $(W\ m^{-2})$, aerosol diffusion S_{dae} $(W\ m^{-2})$ and reflection by the surface S_{dg} $(W\ m^{-2})$. Each term can be estimated by:

$$S_{dr} = 0.79S_0cos(Z)\tau_r\tau_{aa}\tau_{oz}\tau_{aw}\tau_g(1-\tau_r)\frac{0.5}{1-m+m^{1.02}}, \tag{2.24}$$

$$S_{dae} = 0.79S_0cos(Z)\tau_r\tau_a\tau_{oz}\tau_{aw}\tau_g(1-\tau_{ad})\frac{F_c}{1-m+m^{1.02}}, \tag{2.25}$$

$$S_{dg} = (S_{Dn}cos(Z)+S_{dr}+S_{ae})\frac{\alpha_g\alpha_s}{1-\alpha_g\alpha_s}, \tag{2.26}$$

with Z (rad) the sun zenith angle, τ_{aa} and τ_{ad} are respectively the transmittances due to aerosol absorption and scattering, m air mass $(kg\ m^{-3})$ and F_c the ratio of the forward-scattered irradiance to the total scattered irradiance due to aerosols, α_g averaged ground albedo, α_s clear sky or atmospheric albedo.

In this work, following modifications are made by the author. First, proposed by Yang et al. (2001), a different expression for the air mass is used:

$$m = (1-0.0001z_g)/[sin(h)+0.15(57.296h+3.885)^{-1.253}], \tag{2.27}$$

where z_g (m) is ground level and h (rad) is the altitude angle of the sun.

Secondly, in equation 2.24, the expression of the transmittance due to aerosol absorption τ_{aa} is

$$\tau_{aa} = 1-E_1(1-m+m^{1.06})(1-\tau_a), \tag{2.28}$$

with E_1 the constant used in the Bird model associated with aerosol absorption. After testing, the values of $F_c = 0.84$ and $E_1 = 0.1$ as in Bird and Hulstrom (1981) are taken which are more appropriate for the simulations than other suggested values.

Incident Long-Wave Radiation Model

In the simulations presented in this work, the incident long-wave radiation is directly taken from the observed values. In the case where the measurement is not available, $L^{\downarrow}$ $(W\ m^{-2})$ can be generally expressed in terms of the Stefan-Boltzmann law:

$$L^{\downarrow} = \varepsilon_{atm}\sigma T_{atm}^4 = \varepsilon_{clr}F\sigma T_{atm}^4, \tag{2.29}$$

where $\varepsilon_{atm} = \varepsilon_{clr}F$ is referred to as the effective or apparent emissivity and generally varies between roughly 0.7 for clear skies to close to unity for completely overcast skies, with ε_{clr} the clear-sky atmospheric emissivity. F (always ≥ 1) is a cloud factor expressing the increase in clear-sky $L^{\downarrow}$ due to cloud emission, and T_{atm} (K) is the effective atmospheric temperature.

The simple model implemented in *Code_Saturne* is after by Prata (1996): T_{atm} is approximated by the air temperature T_a (K) at reference height close to the surface, and ε_{clr} is estimated by:

$$\varepsilon_{clr} = 1 - [1 + w\ exp(-(1.2 + 3w)^{0.5})] \tag{2.30}$$

where the precipitation water content w equals $46(e_a/T_a)$ with e_a (Pa) the water vapor pressure.

3.4. Surface Temperature Models

The surface temperature T_{sfc} (K) is a key control for energy exchanges at the urban surfaces, particularly for the net long-wave radiation and turbulent energy fluxes. The surface temperature is controlled by the balance between the external net fluxes and the conductive heat flux through the wall, from the building inside to the outside. As a first step in this development, a simple surface temperature model, the force-restore approach (Deardorf, 1978) has previously been implemented in *Code_Saturne*. Because of the limiting hypotheses of this approach (more adapted for the soil model and does not explicitly take into account the conduction) and in order to more accurately calculate the surface temperature of buildings, the author additionally tests in this work a 1D surface temperature approach, called hereafter the wall thermal model. Thus, the surface temperature in the simulations is computed from either the force-restore scheme or the wall thermal scheme. In this section, I briefly describe these two approaches. A comparison of these two surface temperature models with simulation results and observation data is presented in Chapter 3. Actually, a hybrid approach taking the advantages of both the force-restore method and wall thermal scheme has also been tested at the end of this work (cf. Chapter 4 and 5).

The Force-Restore Method

The force-restore approach (Deardorf, 1978) is commonly used in order to calculate the ground temperature in atmospheric models and is considered a very useful tool. In this approach, a prognostic equation for temperature is used to reproduce the response to periodic heating of the soil. The model is based on an analytical solution of a two-layer decomposition of a material considered homogeneous. The deep soil temperature calculation appears in the restore term of the force-restore equation. This model has been extended to urban surfaces (Johnson et al., 1991; Dupont and Mestayer, 2006). Hence, the prediction of the surface temperature is made by using following equation:

$$\frac{\partial T_{sfc}}{\partial t} = \frac{\sqrt{2\omega}}{\mu_t}(L^* + S^* - Q_H) - \omega(T_{sfc} - T_{g/b}) \tag{2.31}$$

where ω (Hz) is the Earth angular frequency, L^* ($W\ m^{-2}$), S^* ($W\ m^{-2}$), Q_H ($W\ m^{-2}$) are respectively net long-wave, net short-wave, and sensible heat flux, μ_t ($J\ m^{-2}\ s^{-0.5}\ K^{-1}$) the thermal admittance and $T_{g/b}$ (K) either deep soil or internal building temperature.

This extension nevertheless supposes well insulated buildings with a nearly constant internal temperature and homogeneous material.

Wall Thermal Model

Neglecting the anthropogenic flux and the latent heat flux terms, the simplified energy balance can read as:

$$Q_{cond} + Q_H = L^* + S^*, \tag{2.32}$$

with Q_{cond} ($W\ m^{-2}$) the conduction heat flux between the building and the underlying ground including net change of energy storage by the building materials and the enclosed air volume.

1D conduction within the patch substrate is bounded by surface energy exchanges at the patch surface and by the internal building energy exchanges or a deep-soil temperature at the substrate base. In the hypothesis of a single layer to express the conduction term, equation 2.32 reads after expression of each term:

$$\frac{\lambda}{e}(T_{sfc} - T_{int}) + h_f(T_{sfc} - T_a) = \varepsilon(L_a + L_e - \sigma T_{sfc}^4) + (1 - \alpha)(S_D + S_f + S_e) \tag{2.33}$$

where λ ($W\ K^{-1}\ m^{-1}$) is the average thermal conductivity of the wall [1], e (m) the thickness of the wall, T_{int} (K) the internal building or deep soil temperature, h_f ($W\ m^{-2}\ K^{-1}$) the heat transfer coefficient calculated from equation 2.36 and T_a the external air temperature (K).

Internal Building Temperature Model

In some cases, such as that where the building is not well insulated; the variation of the internal building temperature can be important and may have a significant impact on the surface temperature (see Chapter 3). In fact, this variation may also depend on building's solarization, levels of inertia and air permeability. Moreover, the internal temperature is rarely measured in the experiment. Its variation may be characterized as compared to the outside surface temperature by three parameters: the mean value, amplitude and phase-change. In order to more accurately model the internal temperature and take into account

[1] The average thermal conductivity is the notion for the one-layer wall model. For a multi-layer wall model, the notion of the equivalent conductivity or conductance ($W\ K^{-1}$) is more appropriate.

its variation, in this work it is computed with an incremental-adjustment method modified after Masson et al. (2002) and similarly used by Krayenhoff and Voogt (2007):

$$T_{int}^{n+1} = T_{int}^{n-1}(\frac{\tau - \Delta t}{\tau}) + \overline{T}(\frac{\Delta t}{\tau}), \tag{2.34}$$

where T_{int}^{n+1} (K) and T_{int}^{n-1} (K) are the computed internal temperatures at the following and previous timestep respectively, Δt is the timestep (s), τ (s) refers to the number of seconds in a day, and $\overline{T}$ (K) is the average over all the outside building surface temperatures computed at timestep n (e.g., MUST case in Chapter 3) or is set one by one surface temperature (e.g., CAPITOUL case in Chapter 4). For a diurnal simulation, the initialized internal temperature value can be considered as the same as the initialized outside surface temperature (e.g., T_{sfc} at midnight) or the outside surface temperature at half an hour before if the observation data is available (e.g., MUST and CAPITOUL cases).

A potential improvement would involve accounting for the full internal building energy balance proposed by Mills (1997) and also used in SOLENE (Miguet and Groleau, 2002), but is not included at present. In order to improve the results, it is possible to replace in Eq.2.34 $\overline{T}$ from the calculation with the average of the measured outside surface temperatures, if available. The latter is included as a user-controlled option in the developed model. A plot of internal temperature modeled by three methods in a building in a diurnal variation

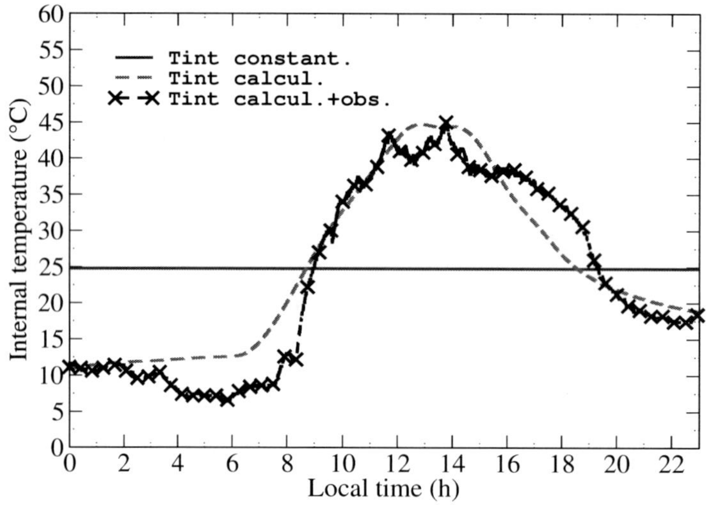

Figure 2.2. Different methods to model the internal building temperature T_{int} (°C) (Tint constant.: taking a constant value; Tint calcul.: using evolution equation; Tint calcul.+obs.: using evolution equation and taking the average outside building surface temperature from the measurements for $\overline{T}$.

profile, this is shown in Figure 2.2. It should be noted that the extrapolation of the internal temperature profile using evolution equation (dashed line in Fig. 2.2) might be different in the simulation because $\overline{T}$ also depends on the resolution of the mesh. It should also be noted that a constant internal building temperature (full line in Fig. 2.2) effectively represents the inclusion of an anthropogenic heating term when the external air is cooler than the interior one (Arnfield, 2003). On the other hand, it is an artificial heat sink when the external air is warmer than the building interior. In fact, the mean building outside surface temperatures are quite sensitive to the internal building temperature evolution, both in terms of magnitude and time-evolution of the diurnal pattern. I will further discuss this point in Chapter 3.

Brightness Temperature

Many comparisons with observations are made in terms of brightness surface temperatures since it is the quantity measured by infrared thermometers (IRT) whilst model-calculated surface temperatures result from the solution of the surface energy balance (see Chapter 4). Brightness surface temperatures account for the surface emissivity and multiple reflections (together, effective emissivity) that are inherent in the observations. Surface brightness temperature (T_{br} (K)) is defined as the temperature that yields to an emitted broadband thermal radiance equivalent to the sum of the true broadband emitted radiance (with reduction due to gray body emissivity) and the broadband reflected radiance (after multiple reflections on canyon surfaces) (Krayenhoff and Voogt, 2007):

$$T_{br} = \sqrt[4]{\varepsilon T_{sfc}^4 + \frac{(1-\varepsilon)L^{\downarrow}}{\sigma}}. \tag{2.35}$$

3.5. Convection Model

The surface convective heat flux (sensible heat flux) must be computed to both solve the surface energy balance (Eq.4.2) and determine the surface-air thermal gradient and therefore turbulent transport. In *Code_Saturne*, the convective heat transfer is computed in 3D for each surface patch. The buildings are explicitly defined in the simulations. Therefore, the detailed representation of the surface allows for a more complex 3D spatial representation of wind speed, turbulence, and temperature than simple canopy averages or vertical profiles. A rough-wall boundary condition is used in this work, based on the logarithmic law modified by the stratification. Usually, these modified laws are based on the Monin-Obukov similarity but are implicit and, therefore, need to be solved iteratively. Here an explicit approach based on the work of Louis (1979) and described in Musson-Genon et al. (2007) is used. The heat transfer coefficient h_f in Eq.4.2 is computed for each solid sub-facet, depending on the local friction velocity u_* $(m\ s^{-1})$:

$$h_f = \frac{\rho C_p u_* \kappa f_h}{\sigma_t ln(\frac{d+z_0}{z_{0_T}})\sqrt{f_m}}, \tag{2.36}$$

where C_p is specific heat $(J\ kg^{-1}\ K^{-1})$, κ is von Kármán constant, σ_t the turbulent Prandtl number, d (m) is the distance of the cell center to the wall, z_0 the roughness length (m), z_{0_T} the thermal roughness length (m), f_m and f_h are the Louis (1979) stability functions which take a value of 1 for neutral conditions. For vertical walls, the neutral conditions are applied.

Chapter 3

Application I: Idealized Urban Areas

In the previous chapter, the new coupled thermal-radiative-dynamic 3D scheme in *Code_Saturne* has been introduced. Now, let us see some applications of the model. Firstly, to ensure modeling accuracy, validating the model results with experimental data is necessary. The chapter shows the first evaluation of the coupling with data from the Mock Urban Setting Test (MUST) field experiment. In MUST, the buildings are idealized by an array of shipping containers. A moderate wind speed day has been simulated by modeling a detailed flow field. Furthermore, on the radiation aspect, the performance of the radiative model will be discussed in comparison to SOLENE model. Then, in order to improve the model, different numerical simulations will be presented for low wind speed and higher building density than in MUST, varying the thermal exchange model and analyzing the differences. On one hand, this assess the thermal impact on the flow field in different conditions. On the other hand, it illustrates the use of the 3D radiative scheme.

1. The Mock Urban Setting Test

1.1. Configuration of the Experiment

The Mock Urban Setting Test (MUST) (Biltoft, 2001; Yee and Biltoft, 2004) conducted in the Utah desert is a near-full-scale experiment that consists of measurements in an idealized urban area represented by 120 shipping containers (length L $\times$ width W $\times$ height H = $12.2 \times 2.42 \times 2.54 m$) arranged in a regular array. MUST has already been used to validate dynamics and dispersion models (Brook et al., 2002; Hanna et al., 2002; Camelli et al., 2005; Milliez and Carissimo, 2007, 2008). Because temperature data are also provided, the MUST field experiment was used to study in detail the thermal-radiative-dynamic coupling (Qu et al., 2011b, 2012a). The study was focused on one instrumented container within the array, and therefore, the computational domain was reduced to three rows of three containers.

For the MUST experiment, the day of 25 September 2001 was selected because this day had already been partly simulated for studies on dispersion (Milliez and Carissimo, 2007) and moreover because a complete 24-h dataset for the upstream wind and the surface temperature was available (which was not the case for other days). During this day (Fig. 3.1), the wind velocity varied from $U_{min} = 3\ m\ s^{-1}$ to $U_{max} = 11.5\ m\ s^{-1}$ and the average diurnal air temperature was about 24°C (measured at $10m$). For this coupling study, the wind speed may be a little high to test strong radiative effect on the airflow, but it emphasizes the convective effects on the surface temperature.

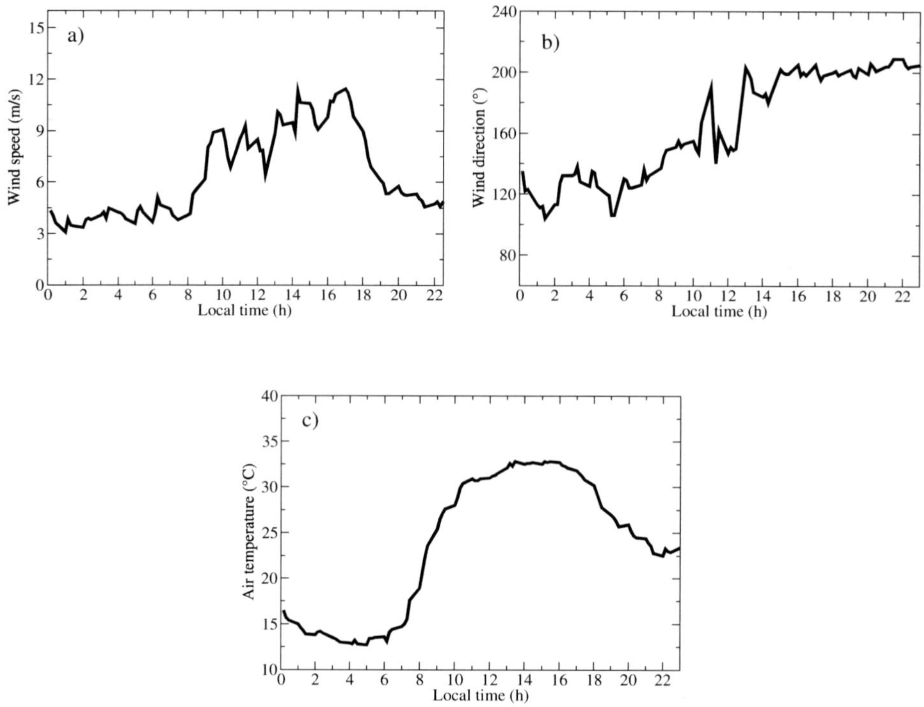

Figure 3.1. Meteorological data at $10m$ from the MUST experiment for 25th of September 2001: (a) wind speed (ms^{-1}), (b) meteorological wind direction (°), and (c) air temperature (°C).

1.2. Simulation Setup

Table 4.1 gives the parameters used in the simulation. To be consistent with the experiment, the wind inlet boundary conditions are determined from measurements, using a meteorological file that contains the wind velocity, turbulence kinetic energy, dissipation rate, and temperature profiles for every $2h$. The variation of the deep-soil temperature is neglected. The internal building temperature is computed by the evolution equation with T from measurements (Eq.2.34). Because some thermal properties were not determined during the experiment, their values were taken from the literature: e.g., the values of the albedo and emissivity of the wall were taken to be those of corrugated iron (Oke, 1987). The thermal admittance was taken assuming an insulating material in the walls, as observed in some pictures. The ground albedo as that was input to the model was evaluated from the incoming and outgoing solar fluxes measured upstream by pyranometers and depends on the zenith angle.

Table 3.1. Parameters employed in the MUST simulations, with Z_{ref} being the reference height for forcing data, U_{ref} being the initial wind speed at Z_{ref}, and ϕ being the initial wind angle. z_0 is the roughness length for momentums. z_{0_T} is the roughness length for heat. α is the short-wave albedo. ε is the long-wave emissivity. μ_S is the admittance.

Parameter	Unit	Value
Dynamic		
Timestep	s	10
Z_{ref}	m	47.5
U_{ref}	$m\ s^{-1}$	3.21
ϕ	$^\circ$	340
Soil z_0	cm	2
Wall z_0 = Roof z_0	cm	0.2
Radiative		
Timestep	min	5
Soil z_{0_T}	cm	0.2
Wall z_{0_T} = Roof z_{0_T}	cm	0.02
α_S		From measurements
$\alpha_W = \alpha_R$		0.1
ε_S		0.8
$\varepsilon_W = \varepsilon_R$		0.13
μ_S	$J\ m^{-2}\ s^{-0.5}\ K^{-1}$	1200
$\mu_W = \mu_R$	$J\ m^{-2}\ s^{-0.5}\ K^{-1}$	1600
Initial temperature values		
T_{air}	°C	16.45
T_S	°C	15.88
T_{Sint}	°C	21.28
T_R	°C	9.28
T_{NW}	°C	10.50
T_{SE}	°C	11.68
T_{NE}	°C	11.87
T_{SW}	°C	12.37
$T_{Wint} = T_{Rint}$	°C	24.28

1.3. Results

Numerical Sensitivity

Before starting the simulations, a numerical sensitivity study has been performed. The evolution of surface temperature with different grids has been tested (e.g., 4 000, 55 000,

173 000 cells). Finally, using the grid with 55 000 cells is a good compromise between precision and computation time.

In the model, the DOM was implemented with two angular discretizations: 32 or 128 directions, which influence the prediction of the diffuse solar flux and the infrared flux. In this case, the results obtained with 32 directions are very close to the ones with 128 directions, but the calculation is faster by a factor of 5. So 32 directions is taken in the remaining simulations.

The timestep for the dynamics was set to $0.1s$ after Milliez and Carissimo (2007). A timestep was introduced for the radiative scheme that is different from the one used for the dynamics. Several radiative timesteps have been tested: 1, 5, 15, and $30min$ and $1h$ for a diurnal cycle simulation. The resulting difference between a timestep of $1min$ and $5min$ is small, being less than 1°C. So $5min$ was considered as an optimum radiative timestep for the simulations.

The values of the surface parameters were taken from the literature. The values for the thermal properties of metal cover a wide range (Oke, 1987), however, a sensitivity study of the variation of the surface temperatures was performed when varying the parameters in the range given by the literature. Table 4.2 illustrates that a change in albedo, emissivity, or admittance, in the range given by the literature, can make a difference of about 1°C $\sim$ 10°C for the surface temperature.

Table 3.2. Deviation of the surface temperatures (°C) during a diurnal cycle by varying albedo, emissivity, and thermal admittance ($J\ m^{-2}\ s^{-0.5}\ K^{-1}$), respectively.

Parameter	Variation	Roof	Northwest	Southeast	Northeast	Southwest
Albedo	+0.1	−2.6	−1.0	−4.5	−1.5	−2.3
Emissivity	+0.1	−1.3	−1.2	−3.5	−1.0	−3.2
Admittance	−400	+1.8	+2.0	+10.1	+2.6	+2.5

The 24-h simulation in parallel computing on a workstation with 8 processors took approximately 4.5 days. A dynamical timestep of $0.5s$ have also been tested, which reduces the computational time to less than 1 day: the results are close to the ones obtained with a dynamical time step of $0.1s$ from 0 to $9h$, but after $10h$ an important difference (12% $\sim$ 30%) appears when the wind speed exceeds 6 $m\ s^{-1}$. In cases in which the wind speed is small, the dynamical timestep could be set up to $0.5s$ to significantly reduce the CPU time.

Simulation of September 25th

Figure 3.2 shows the evolution of modeled and measured surface temperatures using the force-restore method, with coupled thermal-radiative and dynamical model. The diurnal evolutions of the surface temperatures of the roof, southeast face, and northeast face are correctly reproduced by the coupled model. For the northwest face and the southwest face, the simulations show a delay in the morning warming. This delay in warming can be explained by the conduction between the container walls, which contributes to the fast warming in the northwest and southwest faces before they are in the sun but which is neglected in the simulations. In addition, this delay may be inherent to the force-restore method, which overestimates in this case the relaxation to the internal temperature and therefore enhances the thermal inertia. This inertia effect is also observed after sunset for the simulated northwest surface temperature, which shows a delay in cooling. In the afternoon when the atmospheric radiative forcing increases, however, the modeled surface temperatures compare well to the measurements. The results presented above were obtained by

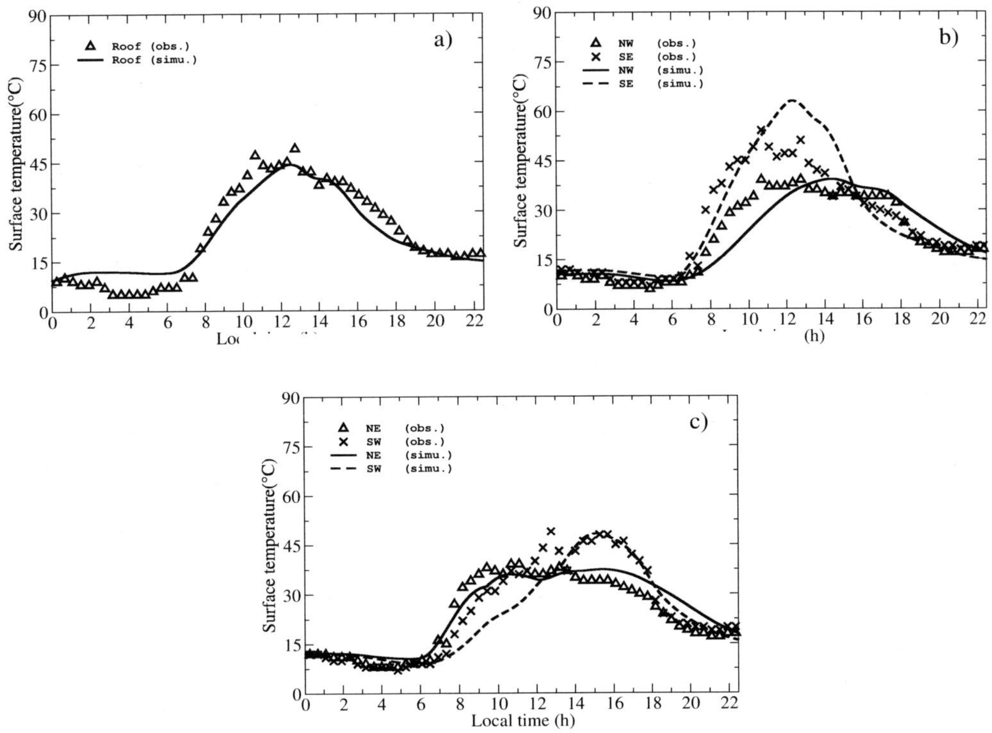

Figure 3.2 Evolution of surface temperature of (a) roof, (b) northwest, southeast, (c) northeast, and southwest faces modeled using the force-restore method during a diurnal cycle.

using the force-restore method and are now compared with those obtained with the wall thermal model (see equation 2.33). The emissivity was chosen after Oke (1987) for corrugated iron. Because some of the containers in the MUST array were painted, a test taking a much higher emissivity (i.e., 0.9) was also made. It results in a decrease in surface temperature. This decrease is greater when using the force-restore method, about 10°C for the southeast wall and 5°C for the rest of the walls. For the wall thermal model, the influence of higher emissivity on the surface temperature is less significant, the difference being less than 2°C for all of the faces. This can explain the fact that the internal temperature has a greater influence than the other surface parameters. The wall thermal model also requires the characteristics of the surface material and the thickness and the thermal conductivity of the wall which were not provided in the data. To improve the comparisons with the observations, the value of the conductivity was adjusted to 6 $W\ K^{-1}\ m^{-1}$, which is not that of pure metal, but may be set by assuming an insulating material in the walls. In Fig.3.3, the evolution of the northwest wall temperature is displayed where a thickness of 10*cm* for the wall is used and it is compared with the observations and that obtained with the force-restore method. In the morning, the wall thermal model (represented by the circles) is able to simulate accurately the increase in the northwest wall surface temperature at 0600 LT, with no delay, as opposed to the force-restore model (represented by the full line). An over-cooling of the surface temperature appears after sunset, however this over-cooling may be explained by a wrong estimation of the internal temperature by the evolution equation, which is highly dependent on the other computed surface temperatures.

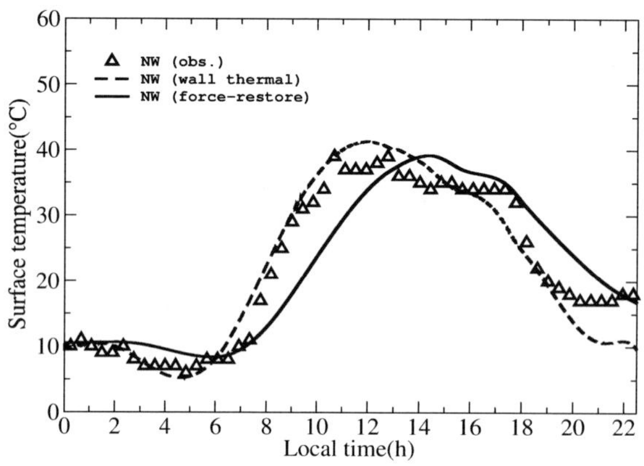

Figure 3.3. The northwest wall temperature evolution obtained with different surface temperature models.

1.4. Summary

New atmospheric radiative and thermal schemes were implemented in the atmospheric module of the *Code_Saturne*. The purpose of this section was to evaluate in detail the coupling between the thermal-radiative scheme and the 3D dynamical model with the field measurements from an idealized urban area, the MUST experiment. The coupled model is able to reproduce the evolution of the surface temperatures for different sides of a container within the MUST canopy during a diurnal cycle despite a delay in warming for the northwest and southwest faces at sunrise. The simulations also showed a significant impact of the convective flux on the surface temperatures. Because the thermal information available in the MUST field is insufficient, sensitivity studies were performed that emphasized the dependence of the model on the parameters describing the building: the properties of the material. In addition, the internal building temperature shows great importance because the buildings are made of metal.

Two ways of computing the surface temperature were compared: the force-restore method and a wall thermal model. Because the force-restore method may be more suited for insulated buildings with a near-constant internal temperature (which is not very representative of the MUST containers), a one-layer wall model has also been tested. Using an appropriate evolution equation for the interior buildings, the force-restore shows good results during the afternoon. It induces a delay in warming at sunset, however, and for the northwest wall a delay in cooling, because of the thermal inertia inherent to the method. The tested one-layer wall thermal model also show some weakness, since it can reproduce the diurnal cycle of the different surface temperature only with a very low thermal conductivity. Nevertheless, it seems more adapted to model shipping containers than is the force-restore method. One proposed improvement is to implement a multilayer 1D thermal model for the walls, as in Masson (2000) or Krayenhoff and Voogt (2007), which may be more adequate for such surfaces. In fact, shipping-container surface temperatures appear difficult to predict with classical urban models, which for instance do not take into account the conduction between the walls, which can be neglected for real buildings but should not be for metal containers, and a 3D conduction wall model may in this case be necessary.

The simulation of realistic atmospheric conditions in the urban areas made possible by this work can be used for various applications. A first example is to study pollutant dispersal in a low-wind case (see Chapter 5). A good description of the heat transfer is essential to describe the convective movement of the air in the streets and is very important for air pollution investigations. A second example is the energy balance of buildings. Estimating the convection fluxes in simple models can keep the computing time low and has application in mesoscale studies; nevertheless, at microscale, it can lead to misleading values in the estimation of the energy loss from the buildings to the atmosphere. In this case, a good prediction of the convective flux can be helpful to the management of the energy consumption and a useful tool in building design.

This study is the first step to validate this thermal-radiative-dynamic coupling model. This 3D modeling investigation can bring more detailed information both on radiative and convective fluxes in very local-scale studies.

2. Comparison of Three Schemes of Increasing Complexity for Predicting Surface Sensible Heat Flux

In this section, sensitivity testing is done to compare three schemes used for predicting surface sensible heat flux. The simulated case is based on the MUST geometry with an upstream wind direction of $-45°$, a reference 10-m wind speed $U_{ref} = 4\ m\ s^{-1}$, and an initial air temperature of 18°C. The simulation starts at 1200 LT for a period of $30min$.

2.1. Constant h_f Model

This scheme is usually used in architecture simulation tools (Miguet and Groleau, 2002; Asawa et al., 2008). The radiative model in this type of tool is very accurate, usually using a detailed 3D geometry. The convective model is very simplified, however, and the scheme considers a constant transfer coefficient. For the comparison, the constant h_f is taken as the average value on each wall from the 3D convection model that the author presented in Chapter 2. Here, h_f is taken equal 14.45 $W\ m^{-2}\ K^{-1}$ for the roof and 6.12 $W\ m^{-2}\ K^{-1}$ for the walls.

In fact, if instead of taking the same constant h_f for all of the walls, separate transfer coefficients are taken for each surface (roof 14.45; northwest face 3.94; southeast face 10.78; northeast face 1.38; southwest face 8.35), it can be better taken into account the orientation of the surface in the wind flow, which decreases the wall surface temperatures by about $2 \sim 4\ K$.

2.2. One-Dimensional h_f Model (1D h_f)

In this model employed in TUF-3D Krayenhoff and Voogt (2007) and similar to the one used in Masson (2000), the transfer coefficient is calculated based on a simple relationship Martilli et al. (2002):

$$h_f = 11.8 + 4.2u(z) - 4.0, \tag{3.1}$$

with $u(z)$ being the vertical wind profile within the canopy. Many authors model this wind profile within the canopy with an exponential law (Cionco, 1965; Rotach, 1995; Krayenhoff and Voogt, 2007). For instance, in TUF-3D, Krayenhoff and Voogt (2007) used an iterative way to find a profile of the exponential form with three coefficients. Here, the vertical velocity is modeled with the exponential profile of Macdonald (2000), which is well adapted to low-density arrays:

$$u(z) = u_H exp[a(z/H - 1)], \tag{3.2}$$

where u_H is the mean velocity at the top of the obstacles and the constant a is the attenuation coefficient, which is determined by fitting the average wind profile within the obstacle array.

2.3. Three-Dimensional h_f Model (3D h_f)

The full model is three-dimensional not only in terms of the radiative exchanges but also the convective exchanges. In this approach, h_f is computed by resolving the 3D RANS and energy equations in the whole fluid domain. Coefficient h_f is calculated for each sub-facet depending on the local friction velocity (Eq.2.36), and the sensible heat flux is calculated with the local air temperature (Eq.4.2).

2.4. Discussion

Figure 3.4 illustrates the effect of the three convective schemes by visualizing, successively, the transfer coefficient, the sensible heat flux, and the surface temperature. The three convective schemes show a difference of the sensible flux of approximately $100 \sim 240\ Wm^{-2}$ for the southeast face and northeast face. With the constant h_f model, the surface temperatures are more homogeneous than in the other two cases. In the MUST configuration, the building array is not dense, and therefore, the effects of the shadow and the multi-reflections are small. That is the reason why the temperatures in the constant h_f approach show little difference within each wall. With the 1D h_f model, we can obviously see the 1D inhomogeneity of the surface temperatures, which is linked to the exponential law wind profile. The 3D h_f model results show the 3D inhomogeneity of the surface temperatures, linked to the inhomogeneity of the 3D wind. On the same face with the same material, we can have a difference of temperature of about 4°C. These results demonstrate the effects of realistically computing the convection fluxes on the surface temperature in the urban areas. Note that in the comparison of the three convective schemes only the transfer coefficient is changed and not the air temperature (which is computed for each grid cell of the fluid domain in three dimensions). A simple air temperature model could lead to additional differences.

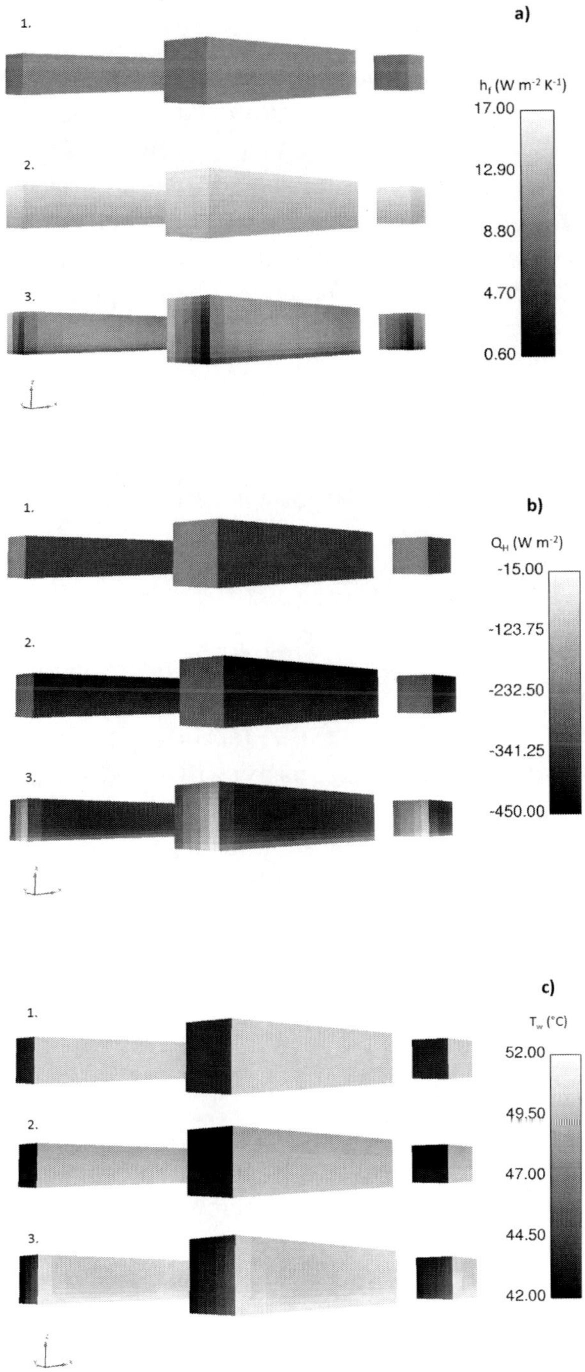

Figure 3.4. Comparison of three convective models with visualization of the southeast and northeast walls at 1230 LT: (a) transfer coefficient h_f ($W\ m^{-2}\ K^{-1}$), (b) sensible heat flux Q_H ($W\ m^2$), (c) surface temperature T (°C). Label 1 indicates the constant h_f model, label 2 is for the 1D h_f model, and label 3 represents the 3D h_f model.

2.5. Summary

In this section, a comparison of different approaches has been made to estimate the convective exchanges at the surface between the developed 3D modeling which consists of solving the 3D RANS equations in the whole fluid domain and two modeling approaches found in the literature: the first approach consists in using a constant heat transfer coefficient, and the second approach is to use a 1D equation based on a vertical wind profile within the canopy. The comparisons are made in terms of the convective transfer coefficients, sensible heat fluxes, and surface temperatures. The three schemes give values of the same order of magnitude for the average surface temperature; nevertheless, only the 3D approach can reproduce the inhomogeneous effect on the wind on a surface: the difference of the same wall can reach 4 °C.

3. Comparison of Two Radiation Models: *Code_Saturne* and SOLENE

3.1. Introduction

In order to more accurately model the physics of an urban canopy, new concepts in surface modeling have been developed. Several canopy models have been mentioned which aim to solve the Surface Energy Balance (SEB) for different applications (cf. Chapter 2). In order to model the radiation exchanges, many SEB models (e.g., Miguet and Groleau (2002); Asawa et al. (2008); Krayenhoff and Voogt (2007)) employ the radiosity method. That is, all radiative reflection and emission is assumed in a transparent medium. The effects of surface morphology on the radiative exchange between a closed system of surfaces surrounding a non-absorbing medium are expressed in terms of the view factors or shape factors for the system of surfaces. Surface morphology affects the radiation balance by altering the magnitude and source of the radiative flux densities incident on each surface. The view factors or shape factors are required when applying the radiosity method to an urban street canyon which are between two equally sized, in-line, parallel planes and between two perpendicular planes sharing a common edge (A_1 and A_2), as shown in Figure 3.5.

Shape factors for the radiative exchange between any two of the canyon surfaces can be derived from one of these two generic types. In general, the radiation is reflected multiple times until the remaining unabsorbed radiation is below a user-defined threshold. Short- and long-wave reflection and absorption are modeled with the same method, but are initialized differently to properly account for shading of direct solar, diffuse solar, and for long-wave emitted from both the sky and the urban surfaces. Each patch requires an albedo and an emissivity as input while outputs for both wavelength ranges include total reflected radiation, total reflected radiation escaping to the sky, and total absorbed radiation by surfaces.

Since the aim of this work is to model radiative exchange in the ABL ($0 \sim 200$ m), for instance, eventually with the smoke or fog between the buildings, the Discrete Ordinate Method has been chosen to solve the radiative transfer equation in the whole fluid domain for *Code_Saturne*. This approach is less precise than the radiosity method to estimate the incoming radiative fluxes on urban surfaces, but the radiation flux can be calculated not only on the facets of the urban landscape but also in each fluid grid cell between the buildings

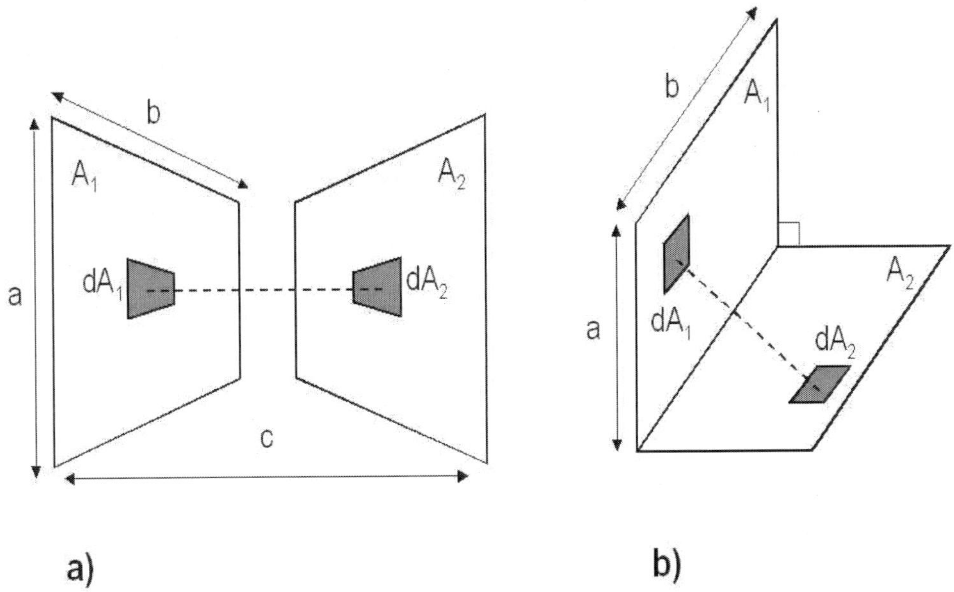

Figure 3.5. Schematics of the view factors or shape factors for the idealized urban street canyons. (a) two parallel in-line planes; (b) two perpendicular planes.

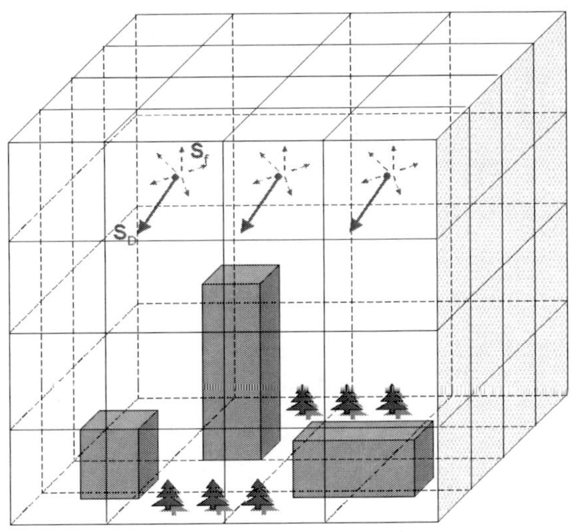

Figure 3.6. Schematics of the radiative model for *Code_Saturne* with: S_D direct solar flux, S_f diffuse solar flux.

in non-transparent medium (Fig. 3.6). Moreover, as previously mentioned (cf. Chapter 2), the mesh is the same in the radiative model and in the CFD model, which makes the thermal-radiative-dynamic coupling computation more simple and convenient.

Therefore, it is interesting to discuss different approaches of the radiative heat transfer models. Here the radiative models in *Code_Saturne* and SOLENE are compared with

MUST experimental dataset.

3.2. Description of SOLENE Model

Initially designed for simulating sunshine, natural lighting and thermal radiation over buildings or urban blocks for architectural projects, SOLENE software is developed by CERMA laboratory ({http://solene.cerma.archi.fr/) and has been applied in many works (e.g., Ringenbach (2004); Bouyer (2009); Hénon (2008) etc.).

Geometry and Mesh

With no particular restriction about the geometry and size, taking into account the urban morphology, SOLENE allows simulating different quantities and phenomena, such as the sunshine features (lighting and mask), the various components of solar radiation (direct and diffuse flux incident on each element with multiple reflections between surface features) and thermal infrared radiation. With the 'contour integration' method (e.g., Soux et al. (2004)) the view factors method implemented in SOLENE may be rapidly and precisely calculated between four-sided surfaces of any orientation.

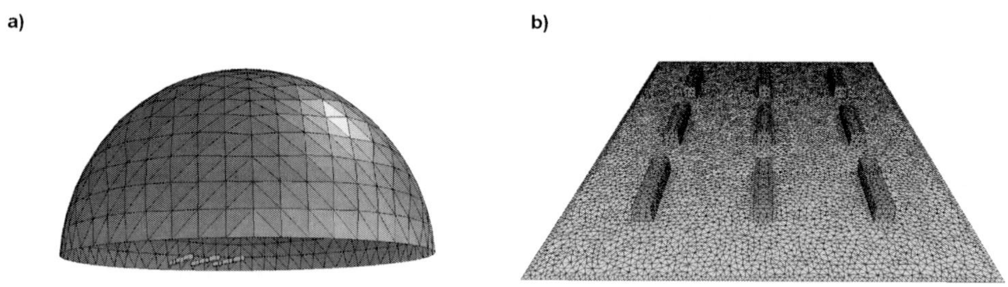

Figure 3.7. Visualization in SOLENE of: a) a dimensionless clear sky hemisphere exhibiting 1024 patches with a radiance distribution at 1500 LST; b) the triangularized shell meshing MUST geometry.

In SOLENE, a geometry is defined by a set of unconnected plane faces, with oriented normal (in general, outward). Each face is constituted itself of one or more plane facets. These facets have known contours, which are polygons defined by the coordinates of each vertex relative to the origin, and by the components of the vector normal. For a same face, the contours are connected and they correspond to the cell of the computational model. Through the 3D geometric modeling of urban morphology, SOLENE triangulates the computational object to a shell mesh. The sky vault is considered as a hemisphere of infinite radius, with the scene to be simulated located at its center. This hemisphere is meshed with a geodesic triangulation of 4^n elements (e.g., most of the time 1024 is a good compromise between precision and speed (Miguet and Groleau, 2002)). The sky is considered as a source of diffuse energy with a non-uniform solar radiation or luminance distribution (e.g., Fig. 3.7a). Different atmosphere and cloud cover conditions are defined by two coefficients, epsilon and delta from a sky model of Perez et al. (1993). Epsilon, ranging from 1 to 12, defines the cloud coverage (more than 6 for clear sky) and delta, ranging from 0.01 to 0.6, defines the opacity linked to the thickness of the clouds, from highly sombre to very shiny. The information assigned to the cell is organized as a descriptor. A descriptor is a set of numerical values assigned to a single cell. Thus, each parameter describing the scene has an associated description file. In addition, for the variables over time, user needs a description file per timestep.

Thermal-Radiative Model

Net Short-Wave Radiative Flux. The direct solar radiation is emitted by the sky in the sun direction while the diffuse solar radiation emission is distributed on all sky facets according to the sky radiance model. In a given scene, SOLENE starts the computation with view factors, incident solar flux and inputs the albedo of each facet. From the incident short-wave radiative flux, the model calculates the total solar flux reflected by the scene and then distributes it to other facets. The contribution of a facet on others is determined by a progressive refinement method: maximal short-wave radiative flux reflected by a facet, then distribute on other facets and determine the new maximal reflected solar flux and so on. The process continues until the difference in the distribution of the solar flux reaches a threshold defined by the user (e.g., 5 W) or other convergence criteria (e.g., the number of iterations).

Net Long-Wave Radiative Flux. In SOLENE, the treatment of the infrared flux is similar to the solar flux by using the progressive radiosity algorithm. The model uses an empirical equation (Monteith and Unsworth, 1990) to estimate the incident infrared flux:

$$L^{\downarrow} = 213 + 5.5T_a, \tag{3.3}$$

where $L^{\downarrow}$ ($W\ m^{-2}$) and T_a (K) are respectively incoming long-wave radiation flux from the atmosphere and air temperature.

However, for the long-wave radiative flux, the facets not only absorb and reflect infrared flux but they also emit. The emission term is determined through the Stefan Boltzmann law $\varepsilon\sigma T_{sfc}^4$ which depends on the surface temperature T_{sfc} (cf. Chapter 2 radiative model). SOLENE can read the atmospheric radiative fluxes and the air temperature meteorological data as input which is the case in this comparison.

Surface Temperature Model. A multi-layers wall model (one or two layers) is implemented in SOLENE to determine temperature, over time, at the surface and inside the wall. Thus, for a single layer wall, the external surface node, with $1/10$ of the heat capacity of the wall, reflects the relatively rapid thermal response of external variations (air, solar and infrared radiation) while the internal node with $9/10$ of the heat capacity plays the inertial role. The two layers wall is built on the same basis.

3.3. Radiation Analyses

Setup for Radiation Computation

The computational effort in time and memory is not comparable. Because the two models are developed with different operating systems (Windows for SOLENE with sequential computing, Linux for *Code_Saturne* with parallel computing). Since no CFD quality criteria needs to be respected for the mesh in SOLENE, the horizontal size from MUST full thermal-radiative-dynamic coupling validation is reduced to save computational time. A standard clear sky model is created with 1024 elements for the SOLENE simulations. A 15 *min* radiative timestep is set both in SOLENE and *Code_Saturne* models. Furthermore, when the multi-reflections are taken into account, 5 iterations are considered for consistency in both models. Based on MUST full thermal-radiative-dynamic coupling validation case (see Table 4.1), the main parameters used in the simulation for the comparison are summarized in Table 3.3. Please keep in mind that all the following comparisons are for the center building in the domain. In addition, hereafter, in order to get great visibility, unless specified otherwise only roof and SE faces are shown in figures. The analysis is the same for other faces.

Comparison of Direct Solar Flux

Let us start with the comparison of direct solar flux before and after multi-reflection by the urban canopy. Through Figures 3.8a, 3.8b, they show that the two models agree quite well with each other. The estimation of the direct solar flux is slightly higher (by less than 10%) for the SOLENE model with and without multi-reflection. These relatively slight visible differences appear only at the south wall (SE and SW faces) and roof which are the most exposed to the sun during the daytime. Moreover, it seems that SOLENE model has a small delay on sunrise. This may be due to the difference on the accuracy of estimating the time of sunrise and the precision of the specified latitude.

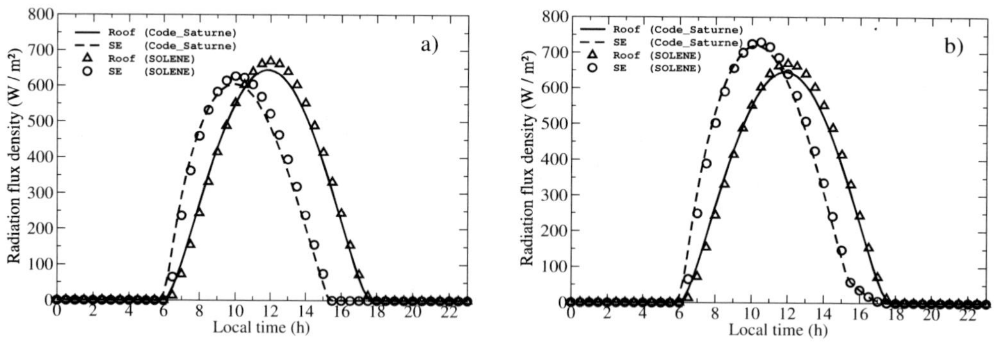

Figure 3.8. Comparison for the central container of the MUST field experiment of direct solar radiation flux on roof and S-E wall during a diurnal cycle between *Code_Saturne* and SOLENE: a) incident direct solar flux; b) direct solar flux after multi-reflections.

Comparison of Diffuse Solar Flux

With regard to the diffuse solar flux, the differences between SOLENE and *Code_Saturne* are significant (Fig. 3.10a), in particular at the four walls, where *Code_Saturne* model predicts the same average diffuse solar flux (superimposed lines in Fig. 3.10a). This can be explained by the fact that the simple clear sky model, Bird Clear Sky model (Bird and Hulstrom, 1981), implemented in *Code_Saturne* assumes the homogeneous and isotropic intensity distribution of diffuse solar flux at inlet boundaries (cf. Chapter 2 radiation model).

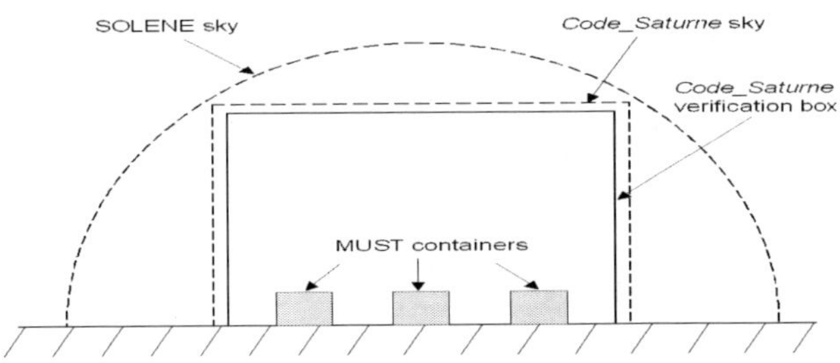

Figure 3.9. Schematics of the procedure to impose the SOLENE values of diffuse solar flux as boundary conditions in *Code_Saturne*. The verification box is used to check the calculated values in *Code_Saturne*.

In order to analyze the difference further the author has introduced the SOLENE heterogeneous diffuse solar flux as the boundary conditions in *Code_Saturne*, using the following procedure (Fig. 3.9):

1. In SOLENE, a 'rectangular building' is created as large as the computational domain of *Code_Saturne* (box with dashed lines in Fig. 3.9) and afterwards, the average diffuse solar flux is computed for each face of this large building using SOLENE.

2. Make a second large 'rectangular building' which is just a little smaller ($-1m$) than the previous one which will be used as verification box for both SOLENE and *Code_Saturne* (box with full lines in Fig. 3.9). Use the average diffuse flux value from the previous step as inlet boundary conditions for each boundary of the computational domain in *Code_Saturne*. Compute the diffuse flux for each face of the second building in SOLENE and *Code_Saturne* (box with full lines in Fig. 3.9). Compare the results to check if the boundary condition is correctly set in *Code_Saturne* (Fig. 3.10b).

3. Apply the diffuse flux value from SOLENE to the MUST reference geometry in *Code_Saturne* (small boxes in Fig. 3.10) and compare the values for the container walls (Fig. 3.10c).

As shown in Figure 3.10b, taking the average diffuse solar flux value from SOLENE as inlet boundary condition gives a very good agreement between SOLENE and *Code_Saturne*. However, the comparison for the container walls is not satisfactory as expected. As shown in Figure 3.10c, the diffuse fluxes of each side of the wall reproduced by

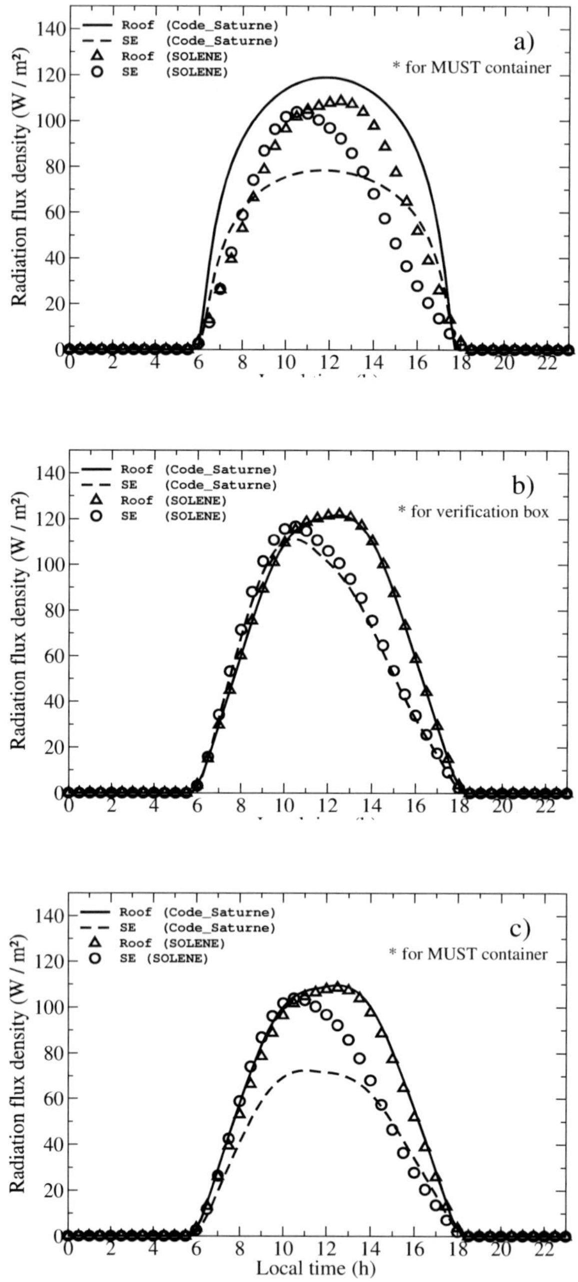

Figure 3.10. Comparison between SOLENE and *Code_Saturne* for incident diffuse solar flux: a) for the container walls with Bird Clear Sky model in *Code_Saturne*; b) using the SOLENE diffuse flux values from SOLENE model as boundary conditions to the *Code_Saturne* for walls of the verification box (see Fig. 3.9); c) for the container walls with SOLENE boundary conditions applied in *Code_Saturne*.

Code_Saturne differ a lot. Only the roof matches well, but all the sides walls exhibit over- or under-estimation. One reason may be the different numerical methods implemented in two models. The developed numerical scheme is diffusive, especially in transparent medium case. Another reason would be that the Discrete Ordinates Method has a quite limited angular at any discretization resolution (32 or 128 directions in *Code_Saturne*). Since the incident diffuse radiation is restricted to propagate in a finite number of directions with an angular sector width (128 directions in this case) from the boundary into the domain, they may not all reach the vertical building surfaces but somewhere else in the domain, except for the roof which is a horizontal surface and at the highest level can receive the diffuse flux from all the directions. This is also proved by the result shown in Figure 3.10b where the distance from the boundary to building surface is very short. Since the direct solar flux is unidirectional, the distance between the building and the boundary does not have the same influence on the results in *Code_Saturne* (Fig. 3.8).

Nevertheless, using the diffuse flux value from SOLENE in *Code_Saturne* still shows an improvement. Moreover, since the diffuse solar flux value is much smaller than the direct flux in clear sky condition, it plays a minor role in the surface energy balance.

Comparison of Long-Wave Radiation Flux

Since the observation of the 180° incident long-wave radiation flux is available in the MUST data, it is used as sky boundary conditions for both *Code_Saturne* and SOLENE. Moreover, the long-wave radiative emission depends on the surface temperature (Eq. 1.6) which introduces more complexities (see next section), here only the output of the average incident long-wave radiation flux at each face of the building for the two models is compared. Figure 3.11 indicates that the prediction of the incident long-wave radiation agrees very well with the two models (dashed line and solid line are superimposed). The average incident long-wave radiation flux on the four sides of building wall exhibits an isotropic behavior (solid/dashed lines are all superimposed and solid lines are very close to dashed lines). Unlike the distribution of the diffuse solar flux which depends on the position of the sun, the distribution of the incident infrared flux at each boundary is isotropic. Therefore the symmetric building geometry makes each wall receive the same quantity of the average infrared flux. Thus, *Code_Saturne* calculates the incident infrared radiation as well as SOLENE.

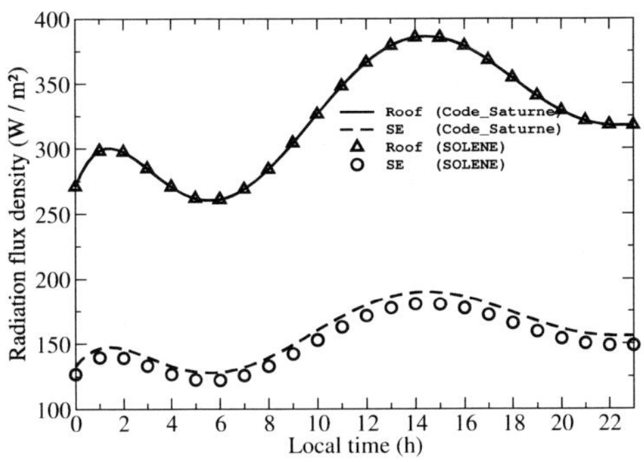

Figure 3.11. Same as Fig. 3.8 but for incident long-wave radiation flux.

Comparison of Surface Temperatures

Up to now, the atmospheric radiation flux have been compared term by term. Since the primary aim of the developed thermal-radiative model is to calculate the exterior surface temperatures of the facets, finally let us look at the prediction of the surface temperatures by the two models. Focusing on radiative aspects, the computation of the sensible heat flux is strongly idealized for this test. The simulations are based on a given constant h_f for each face of the buildings and ground every two hours. The constant h_f values are obtained by averaging the result of the thermal-radiative-dynamic full coupling model in moderate wind condition. Thus, no conservation equations for mass, momentum, and energy need to be solved and, therefore, the influence of a model other than the radiation model can be excluded. The wall thermal model is used from *Code_Saturne* (cf. Chapter 2) and one-layer wall from SOLENE is considered to calculate the surface temperature. The air temperature and incident long-wave radiation are directly imposed from meteorological data. Since the influence of the internal building temperature on the external surface temperature is quite important in the MUST case, the diurnal variation of the interior temperature is updated every half hour. Furthermore, the values for the specific heat capacity C_p and the density ρ in SOLENE (Tab. 3.3) are estimated which are close to the MUST experimental field. In the SOLENE model, the internal building energy balance and the internal transfer coefficient h_{int} must be set. In order to keep an equivalent heating configuration, h_{int} is set to 1000.

Figure 3.12 displays the evolution of modeled surface temperatures overtime using the *Code_Saturne* and SOLENE, and the observation data as reference. As mentioned above, the h_f is only changed every two hours for this moderate wind speed case, this simplification neglects the variations of the wind. As a consequence, the corresponding surface temperature profile may change sharply. This undesired behavior is more remarkable from the simulation of *Code_Saturne* (e.g., NE face in Fig. 3.12c). Overall, the mean surface temperatures are rather well reproduced by both models. The diurnal cycle is also well simulated and the variability of the two models and the observations are comparable. On the other hand, this good result is also consistent with the ability of the full thermal-radiative-dynamic coupling to estimate good average h_f values.

The surface temperature is better simulated by *Code_Saturne* with a bias smaller than 5°C. This may be due to a lack of information on some thermal properties in the experiment. The estimated parameters in the simulation might be better adapted for the one-layer wall thermal model in *Code_Saturne*. Additionally unknown parameters (C_p and ρ) used in SOLENE may lead to more uncertainties. However, it can be seen that SOLENE overestimates the surface temperatures starting from the middle of the day. This behavior of the model is probably associated with a strong contribution of the thermal inertia leading to a delay in warming with storage in the morning then heat releases from the middle of the day.

In addition, both *Code_Saturne* and SOLENE can produce a visualization of the 3D field. For instance, the distribution of the surface temperatures at 1000 LST is pictured in Figure 3.13. The visualization of the building temperatures is consistent with the above discussion. Moreover, the ground temperature reproduced by SOLENE (Fig. 3.13b) is higher than by *Code_Saturne* (Fig.3.13b) ($\sim>$ 5°C). The degradation of the shadow seems more observable from *Code_Saturne* than from SOLENE. This is due to the chosen scale and by recalling that the scheme in *Code_Saturne* has an additional numerical diffusion.

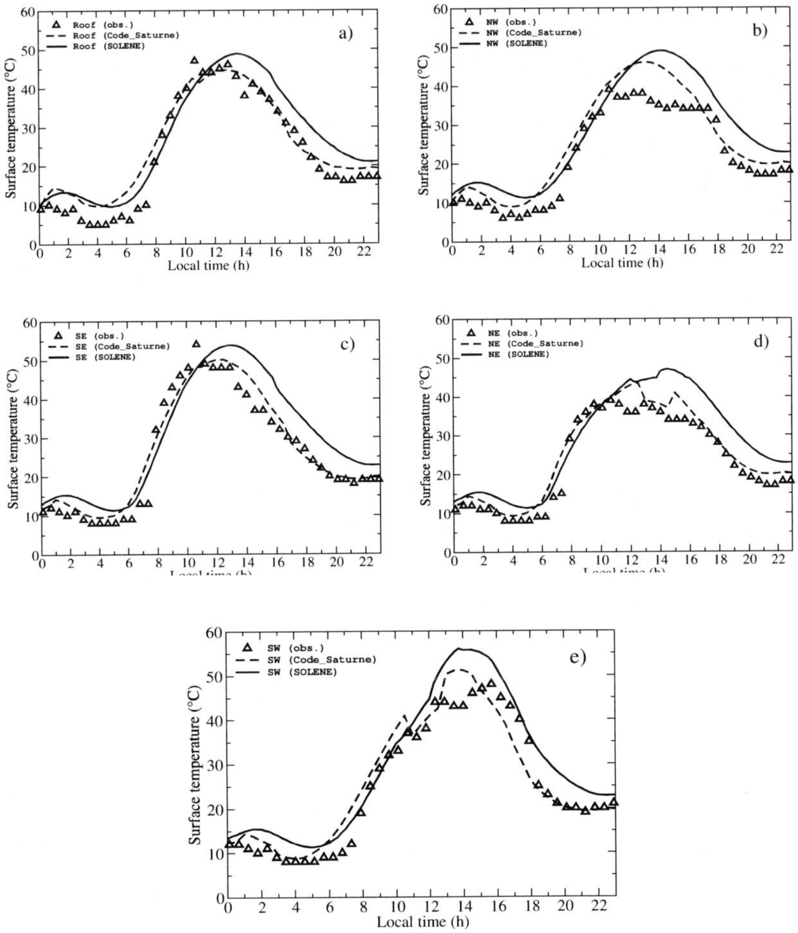

Figure 3.12. Evolution of surface temperature of different faces of the central container in MUST field experiment during a diurnal cycle.

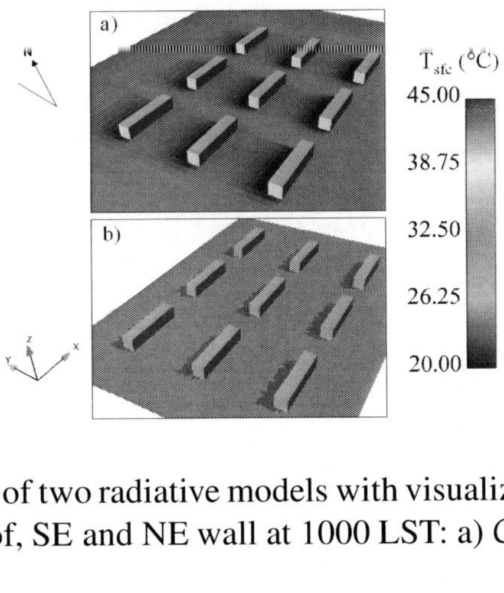

Figure 3.13. Comparison of two radiative models with visualization of surface temperatures (T_{sfc}) (°C) of ground, roof, SE and NE wall at 1000 LST: a) *Code_Saturne*; b) SOLENE.

Difference with the MUST shipping container surface temperatures is highly dependent on the evolution of the internal temperature, but the diurnal variation of the deep ground temperature should be insignificant. So one reason for the ground to appear warmer in SOLENE may be the estimation of the value of C_p and ρ for the soil. By using one-layer wall scheme, they may be sensitive to the solution of the ground temperature in SOLENE for this particular case.

3.4. Summary

In this section, the atmospheric radiation fluxes and the container walls surface temperature simulated by *Code_Saturne* and SOLENE models have been compared with the MUST experimental dataset. It has been found that both models give similar mean daily cycle of radiation fluxes. For the direct solar flux and incident infrared flux, *Code_Saturne* is in very good correspondence with the SOLENE model. The prediction of the diffuse solar flux by the Discrete Ordinate Method of *Code_Saturne* is poorer, due to an assumption of isotropy uniform sky and the different characteristics of the numerical method. One perspective is to implement a similar sky model as SOLENE in *Code_Saturne* to take into account the non-uniformity of the distribution of the diffuse solar flux at inlet boundaries. The diffuse flux reproduced by *Code_Saturne* in this case has already used the higher angular resolution (128 directions) in the current model. It may be worth to test by increasing much more the angular resolution in the model to get some improvements.

Concerning the surface temperature, with imposed heat transfer coefficients, the average diurnal cycle of the surface temperature of different building sides is well reproduced by both models. The difference between two models is more significant from the afternoon. For this reason, the comparison of the emission and reflection of the IR flux reproduced by two models are not discussed because of high dependence on the surface temperature. Incertitude related to additional thermal parameters, C_p and ρ, is more pronounced for the ground temperature in SOLENE. Both of them have facility to the visualization of the 3D field. They can be useful tools for simulating the outdoor thermal environment.

Furthermore, only one-layer wall thermal model is tested here. Using a multi-layer scheme and including C_p and ρ with more detailed information on properties of the material may be appropriate to compare SOLENE and *Code_Saturne*. In cases where the diffuse flux data is not available or in cloudy sky weather conditions, the calculation of the diffuse flux value from SOLENE with a domain size box then to set as inlet boundary conditions leads to improvements in *Code_Saturne* results. Conversely, using a CFD code to more accurately model the h_f is helpful to SOLENE for predicting the sensible heat flux under moderate wind speed condition.

4. Numerical Study of the Thermal Effects of Buildings on Low-Speed Airflow

4.1. Introduction

The influence, on the airflow pattern, of the incident atmospheric radiation on the canyon walls and ground surfaces under conditions of low wind speed can be important and should

Table 3.3. Parameters used in the comparison between *Code_Saturne* and SOLENE for MUST simulations; α is the short-wave albedo; ε is the long-wave emissivity; λ is the thermal conductivity; e is the thickness; T_{air} the initial air temperature; T_S and T_{Sint} are respectively the initial soil temperature and deep-soil temperature; T_{Roof}, T_{NW}, T_{SE}, T_{NE} and T_{SW} are initial individual surface temperature; T_{Wint} and T_{Rint} are respectively building and roof internal temperatures; h_f is the heat transfer coefficient; C_p is the specific heat capacity; ρ is the density; h_{int} is the internal transfer coefficient.

Parameter	Unit	Value
Time-step	*min*	15
α_S		0.55
$\alpha_W = \alpha_R$		0.1
ε_S		0.8
$\varepsilon_W = \varepsilon_R$		0.13
$\lambda_W = \lambda_R$	$W\,K^{-1}\,m^{-1}$	6
λ_S	$W\,K^{-1}\,m^{-1}$	0.75
Wall e = Roof e = Soil e	m	0.1
Initial values		
T_{air}	°C	16.45
T_S	°C	15.88
T_{Sint}	°C	24.28
T_{Roof}	°C	9.28
T_{NW}	°C	10.50
T_{SE}	°C	11.68
T_{NE}	°C	11.87
T_{SW}	°C	12.37
$T_{Wint} = T_{Rint}$	°C	11.11
Top h_f	$W\,m^{-2}\,K^{-1}$	25.32
NW h_f	$W\,m^{-2}\,K^{-1}$	10.08
SE h_f	$W\,m^{-2}\,K^{-1}$	10.83
NE h_f	$W\,m^{-2}\,K^{-1}$	20.29
SW h_f	$W\,m^{-2}\,K^{-1}$	17.79
Soil h_f	$W\,m^{-2}\,K^{-1}$	51.76
Additional values for SOLENE		
Wall C_p = Roof C_p	$J\,m^{-3}\,K^{-1}$	650
Soil C_p	$J\,m^{-3}\,K^{-1}$	800
Wall ρ = Roof ρ	$kg\,m^{-3}$	7000
Soil ρ	$kg\,m^{-3}$	1600
h_{int}	$W\,m^{-2}\,K^{-1}$	1000

not be neglected in pollutant dispersion or urban design studies. Solar heating of the building walls and the ground during the day warms the air near the wall or above the ground, which can cause strong thermally induced air motion. This thermal effects on the airflow, including building geometry and architecture as well as street canyon dimensions, have been mainly studied with wind-tunnel experiments (Uehara et al., 2000; Kovar-Panskus et al., 2002; Richards et al., 2006) and numerical models (Sini et al., 1996; Kim and Baik, 2001; Louka et al., 2001; Xie et al., 2007) and fewer with full-scale or near full-scale experiments (Louka et al., 2001; Kanda, 2006; Niachou et al., 2008). In general, field and wind-tunnel experimentalists stress the technical difficulties in controlling air and surface temperatures and various air or surface temperatures gradients. Moreover, in order to obtain significant thermal effects and respect the similarity with a real building, a very high surface temperature (100 $\sim$ 400 °C) has to be applied in wind-tunnel experiments (Uehara et al., 2000; Kovar-Panskus et al., 2002; Richards et al., 2006).

To overcome this limitation, numerical studies were performed. Sini et al. (1996) first numerically showed that the thermal forcing could markedly influence the air motions and the wind structure. Later, Kim and Baik (2001) simulated the flow in an urban street canyon with various aspect ratios and with street bottom heating using a two-dimensional (2D) CFD model with a $k-\varepsilon$ turbulence closure. Kovar-Panskus et al. (2002) identified the differences in the airflow regimes between the cases with and without wall heating. Xie et al. (2007) investigated the influence of the multi-surface heating configuration on the flow field and pollutant transport problems with the wind-tunnel experiment of Uehara et al. (2000).

However, most of previous research works considered a 2D computational domain and assumed idealized sunlit wall configurations (uniform sunlight on one surface of the canyon and no sunlight on any other solid surfaces) without including a realistic radiative model. Further research work is therefore necessary to improve our knowledge about the thermal effects on the local atmospheric environment.

On the basis of this background, the aim of this section is to assess the thermal impact on the flow field in different thermal conditions by extending the work of the first section of this chapter to lower wind speed and a higher building density than in MUST (Qu et al., 2011c,a, 2012b).

4.2. Simulation Setup

Initial and Boundary Conditions

The computational domain is three-dimensional and the boundaries are positioned far or high enough from the built-up area of interest ($10H$ as shown in Fig. 3.14a). A Dirichlet boundary condition at the inlet persists throughout the simulation (East in Fig. 3.14a). For the top boundary, a constant shear stress is applied, corresponding to the inflow profiles in order to prevent a horizontal change from the inflow profiles. As frequently done in CFD codes, the symmetry boundary conditions are used at the lateral boundaries (North and South in Fig. 3.14a) since the approach inflow direction is parallel to them. At the outlet boundary (West in Fig. 3.14a), the free outflow condition is applied. Thus, the derivatives of all flow variables are forced to zero.

The buildings have the same dimensions ($L \times W \times H$: $12.2 \times 2.42 \times 2.54$ m) as the MUST experiment shipping containers (cf. Chapter 3, Section 1) and the orientation of the x axis is pointing towards the geographic north. In addition, the aspect ratio (H/W, where H is the building height and W is the along-wind spacing) is modified to 1 in order to

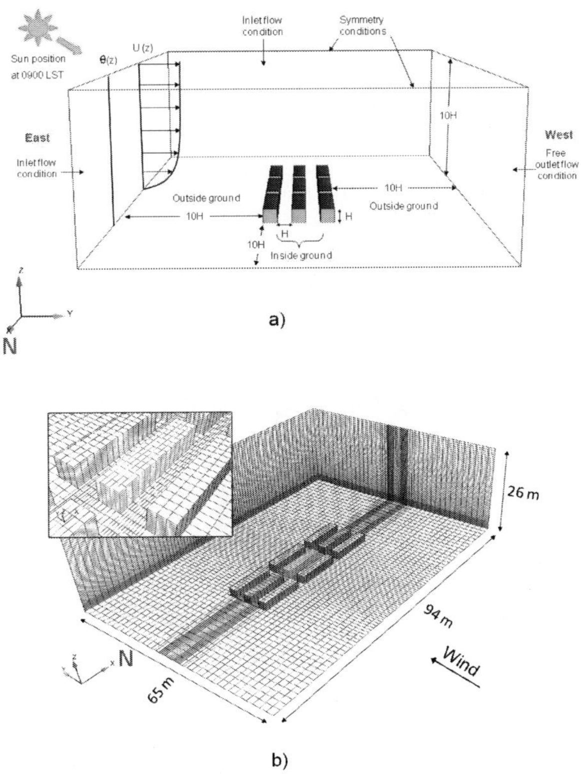

Figure 3.14. Computational domain with building models for low wind speed: a) simulations-definition of boundary conditions; b) mesh of the domain and sub-domain. The north is indicated.

increase the density of the urban canopy and therefore to emphasize the thermal impact on the wind. The run takes place from 0900 LST to 0910 LST. At this time, the Sun position is on the East side of the domain as shown Figure 3.14a. The meteorological initial and inlet boundary conditions are a neutrally stratified logarithmic wind profile blowing westward, associated with a neutral potential temperature profile of 30 °C with both dynamical and thermal roughness lengths set to 4 *cm*. Since the thermal effects are expected to be of substantial importance for low wind speeds, certainly below 3 $m\ s^{-1}$ when the urban dome regime dominates (Oke, 1987; Kovar-Panskus et al., 2002), the 10 *m* high wind speed is set to 1 $m\ s^{-1}$. Low wind conditions and large surface-air temperature differences cause large thermal effects. Different differential temperatures between the heated surfaces and the air were simulated in numerical studies (5 °C in Sini et al. (1996), 10 °C in Xie et al. (2007), 32 °C between the windward wall and the air, 53 °C between the leeward wall and the air in Onomura et al. (2009)). In general, a surface-air temperature difference of 5 to 10 °C is considered to be close to the one of a wall heated by solar radiation. Thus, in this study, the initial air, soil, and wall/roof temperatures are set to 30, 32 and 36 °C respectively.

Table 3.4. Boundary conditions and parameters used in the low wind speed simulations.

Domain dimensions	$52(x) \times 93.68(y) \times 26.54(z)\ m$
Unstructured hexahedral grid resolution	Total 253,943 cells
	$x_{min} = 0.81\ m$, $x_{max} = 1.90\ m$
	$y_{min} = 0.48\ m$, $y_{max} = 1.90\ m$
	$z_{min} = 0.23\ m$, $z_{max} = 0.68\ m$
Inlet boundary conditions	Timestep 0.05 s, Reference height $Z_{ref} = 10\ m$
	Initial wind speed $U_{ref} = 1.0\ m\ s^{-1}$, wind direction relative to the north $-90°$
	$k = 1.88 \times 10^{-2}\ m^2\ s^{-2}$, $\varepsilon = 1.01 \times 10^{-4}$
	Dynamical roughness for soil $z_0 = 4\ cm$, for wall and roof $z_0 = 2\ cm$
Thermal boundary conditions	Thermal roughness z_{0_T} for soil, wall and roof as same as dynamical roughness z_0
	Initial potential temperature $\theta = 30°C$
For the imposed temperatures case	Outside / Inside ground temperature $T_{sfc} = 35.15/11.30\ °C$
For both the imposed temperatures and the radiative transfer cases	Initial walls and roofs temperature $T_{sfc} = 36\ °C$
Additional boundary conditions for the radiative transfer cases	Radiative timestep 120 s
For the soil	Albedo $\alpha = 0.53$, emissivity $\varepsilon = 0.8$
	Conductivity $\lambda = 0.75\ W\ K^{-1}\ m^{-1}$, thickness $e = 50\ cm$
	Deep-soil temperature $T_{int} = 25\ °C$
For the walls and roofs	Albedo $\alpha = 0.1$, emissivity $\varepsilon = 0.13$
	Conductivity $\lambda = 6\ W\ K^{-1}\ m^{-1}$, thickness $e = 10\ cm$
	Internal temperature $T_{int} = 32\ °C$

In order to study the contribution of atmospheric radiation on the airflow, three cases have been simulated: a reference case without heat transfer, i.e., with a neutral stratification of the atmosphere - called hereafter 'neutral case', a case with heat transfer but where the surface temperatures are imposed - called hereafter 'imposed temperatures case', and a case with the 3D atmospheric radiative model coupled to the dynamical one - called hereafter 'radiative transfer case'.

For the second case, the imposed wall/roof temperature is set to the one used to initialize the surface temperatures in the radiative transfer case (36 °C). For the ground, two imposed temperatures are used, which are obtained by averaging the results of the radiative transfer case: one (lower) for the ground inside the canyons (11.30 °C) (in order to take into account the shading effects) and another one for the area outside the array (35.15 °C).

In the two last cases, a buffer zone is included to remove any numerical perturbation due to the difference between the ground-level temperature of the inlet meteorological profile and the imposed or initial ground temperature inside the domain. The numerical simulation features, including initial and wall boundary conditions are summarized in Table 3.4.

Grid Resolution

The grid with sufficient resolution plays a direct role in the quality of the simulations. Based on the unstructured hexahedral grid used in MUST case, the grid resolution is increased in this work as shown in Figure 3.14b. The minimum grid resolution is set to $1/10$ of the building height scale (0.25 m) within the regions of high gradients. In order to keep the truncation error small, an expansion ratio of 1.2 is set for the grid stretching/compression. Thus, the mesh generated in this work has total about 250 000 cells (Tab. 3.4). After a sensitivity study, the timesteps for the dynamical model and the radiative one were respectively set to 0.05 s and 2 min.

4.3. Results and Discussion

The numerical simulation results are analyzed at $t = 600s$ when the mean flow in the entire simulated urban canopy reaches a quasi-steady state. Figure 3.15 presents at 0900 LST the direct solar flux received by the surfaces from the radiative model. It can be observed that the distribution of the shadow in the canyons is mainly behind the leeward walls and the most heated walls are at windward sides.

Comparison of Three Thermal Conditions: Neutral, Imposed Temperatures and 3D Radiative Transfer Cases

Figure 3.16a and b show the mean Turbulent Kinetic Energy (TKE) fields at roof level and half building height level for three thermal conditions. In all conditions, due to the presence of the buildings creating mechanically induced turbulence, more TKE is found at roof level than in the canopy. Compared to the neutral case (Fig. 3.16a1, b), the presence of heat transfer (Fig. 3.16 a2, a3, b2, b3) enhances the TKE in the whole domain due to a large thermal production, in particular near the windward side of the first array (the most heated wall) and in the wake zone (accumulated thermal production). Similar TKE distributions in the street canyons are found in the two cases with heat transfer (Fig. 3.16a2, a3, b2,

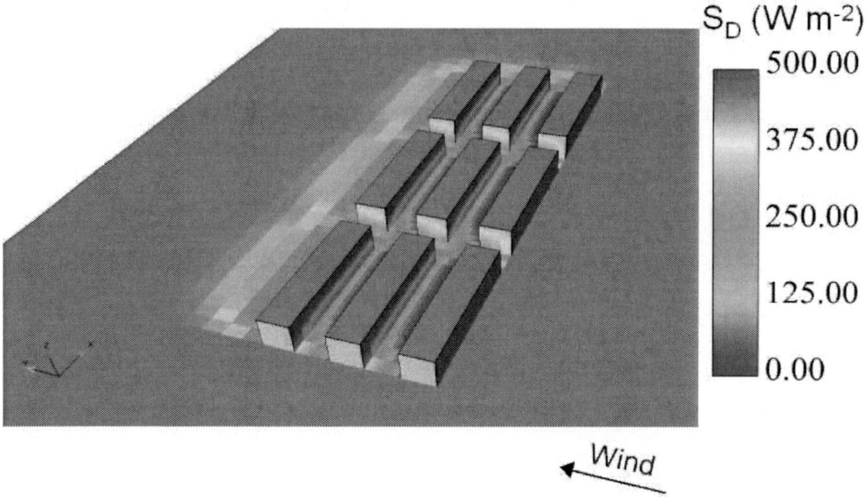

Figure 3.15. Visualization of the direct solar flux (S_D) ($W\ m^{-2}$) received by the surfaces with a 0900 LST sun position from the radiative model.

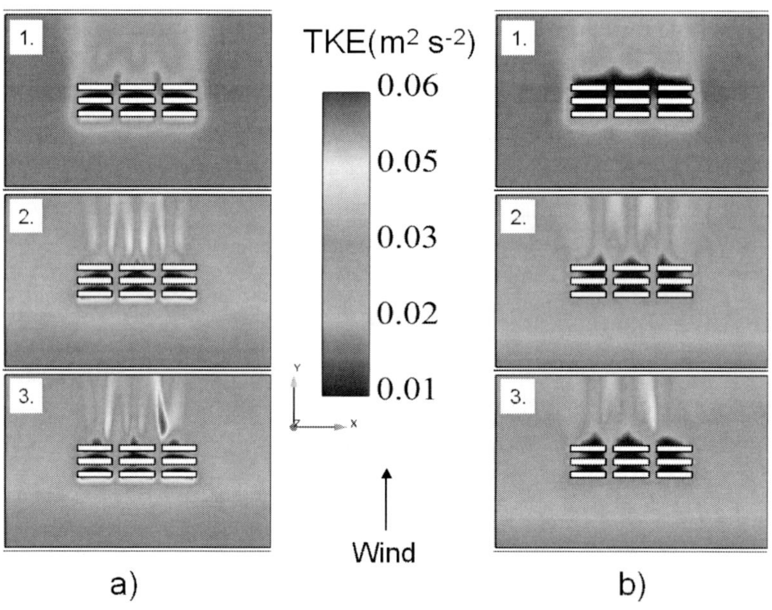

Figure 3.16. Comparison of the mean turbulent kinetic energy (TKE) ($m^2\ s^{-2}$) distribution on: a) roof level; b) half height of the building, for three thermal conditions: 1) neutral; 2) imposed temperatures; 3) 3D radiative transfer with a 0900 LST sun position.

b3) due to surface temperatures on the same order of magnitude, generating similar thermal production. However, the non-uniform surface heating in the radiative transfer case shows an asymmetry in the TKE distribution. This asymmetry obviously shows smaller values in the center and larger values on the right side of the wake zone. The asymmetrical TKE

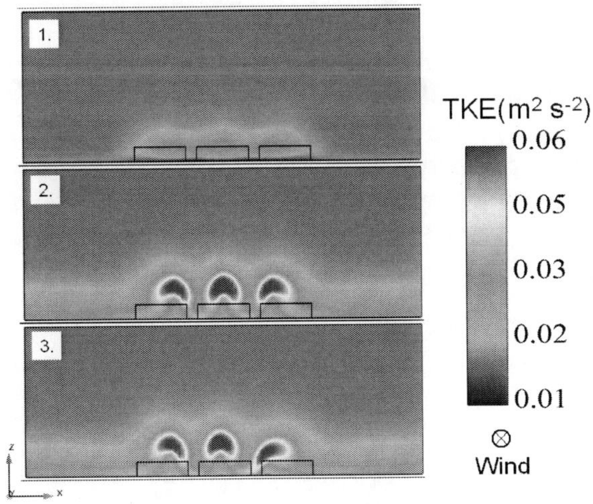

Figure 3.17. Comparison of the mean turbulent kinetic energy (TKE) ($m^2\ s^{-2}$) distribution on the cross-section at $5H$ behind last building for three thermal conditions: 1) neutral; 2) imposed temperatures; 3) 3D radiative transfer with a 0900 LST sun position.

distribution and the larger TKE in the wake zone can be better understood by looking at the dominating vortex structure in the cross-section $5H$ behind the array, as shown in Figure 3.17. Consistent with Figure 3.16a, three different vortex regimes are found above the roof (Fig. 3.16), with larger TKE values in the case with heat transfer (Fig. 3.17.2 and Fig. 3.17.3). In comparison to neutral and imposed temperatures cases (Fig. 3.16.1 and Fig. 3.17.2), involving shadowing and radiation trapping effects in the radiative transfer case induces a different TKE contour in the right side part of the wake zone (Fig. 3.17.3). As mentioned in Chapter 2, the model accounts the thermal stratification, thus the motion

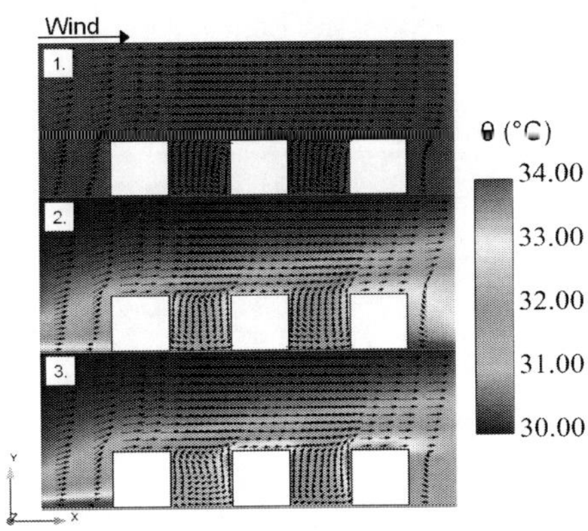

Figure 3.18. Comparison of the mean airflow potential temperature (θ) (°C) distribution on the center-plane for three thermal conditions: 1) neutral; 2) imposed temperatures; 3) 3D radiative transfer with a 0900 LST sun position.

in the canyon induced by thermal effects is substantially vertical. Figure 3.18 compares the distribution of the mean airflow potential temperature and velocity vector maps for the three cases. Velocity vectors are normalized by the reference velocity (i.e., 1 $m\ s^{-1}$) and projected onto the center-plane. Without surface heating (neutral case, Fig. 3.18.1), due to the building density ($H/W = 1$), the airflow pattern in the canopy is a classic skimming flow regime as described by Oke (1987): a clockwise wind flow structure develops in the center of the street canyon. Taking into account surface heating (either imposed temperatures or radiative transfer conditions), the potential temperature is increased mainly at the roof level and windward wall (Fig. 3.18.2 and Fig. 3.18.3). As a result, the airflow structure is characterized by a significant air motion along the windward faces in the cavities. As mentioned previously, the initial ground temperature outside the canyon in the imposed temperatures case is set to a constant value (35.15°C) estimated by averaging the results of the radiative transfer computation. In fact, in the radiative transfer case, the ground temperature close to the first building is higher (+2°C) as a result of short- and long-wave reflections and long-wave emissions between the soil and the first windward face. Behind the last building, the soil temperature in the shaded area is much lower in the radiative transfer case (12 $\sim$ 24°C not shown here) than in the imposed temperatures case (35.15°C), which results in a significant difference in potential temperature.

It should be pointed out that the velocity vectors (Fig. 3.18) show no effect of separation over the forward corner of the upwind building. These flows are always present in shear flow over bluff bodies attached to a wall. In fact, the absence of such structures on the roof is due to the fact that $k-\varepsilon$ model tends to incorrectly predict the balance between production and dissipation of turbulence in strong velocity gradient regions such as the upwind corner of buildings. Removing this negative effect on the convective heat transfer and radiation balances over the structures is still an ongoing work.

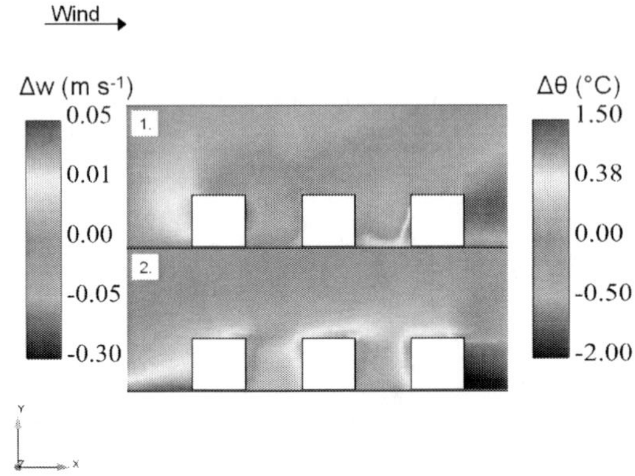

Figure 3.19. Difference between 3D radiative transfer with a 0900 LST sun position and imposed temperatures case on the center-plane: 1) difference in vertical velocity (Δw) ($m\ s^{-1}$); 2) difference in potential temperature ($\Delta\theta$) (°C).

To further investigate the contribution of the radiative model, Figure 3.19 shows the vertical velocity difference (Fig. 3.19.1) and the potential temperature difference (Fig. 3.19.2) in the center-plane between the radiative transfer case and the imposed temperatures case. It should be noticed that taking into account the non-uniform surface heating by 3D atmospheric radiation transfer can substantially modify the distribution of the vertical velocity and potential temperature patterns in the streets, in comparison to simply setting constant surface temperatures: close to the leeward sides, the differences are between -0.3 and $0.05\ m\ s^{-1}$ for the vertical airflow pattern, and between -2 and 1.5°C for the potential temperature in this case (Fig. 3.19).

Comparison of the Different Radiative Transfer Conditions

Up to now, we have seen that the difference between the imposed temperatures and the radiative transfer case is already complex. In fact, depending on the position of the sun and the shadows, the walls and the soil are not heated uniformly, modifying the stratification of the flow and hence the buoyancy forces and turbulence production. Furthermore, even in the same sunlit configuration (i.e., same orientation, same time of the day), the results can be different depending on the values of the parameters used to describe the material physical properties of the building walls or the ground; for instance, the albedo, which may be increased when painting the surface with a light color. In Figure 3.20, using the ra-

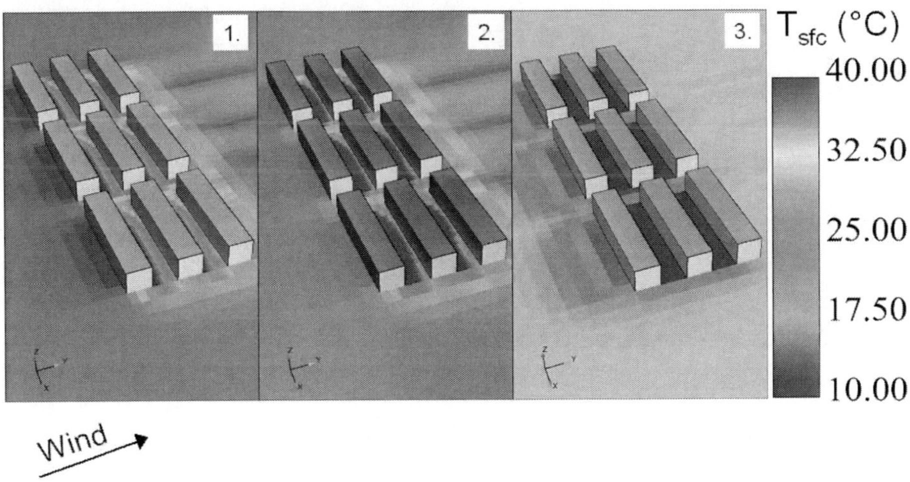

Figure 3.20. Comparison of the surface temperatures (T_{sfc}) (°C) for three radiative transfer conditions: 1) a 0900 LST sun position with an albedo of 0.6 for the building walls; 2) same as 1) but with an albedo of 0.1 for the building wall; 3) same as 2) but with a 1600 LST sun position.

diative heat transfer model, the surface temperatures for three radiative transfer conditions are presented: a 0900 LST sun position with a higher albedo value (0.6 instead of 0.1) for the building walls (Fig. 3.20.1), a reference case same as Fig. 3.20.1 but with an albedo of 0.1 for the building walls (Fig. 3.20.2), and same as Fig. 3.20.2 but for a 1600 LST sun position (Fig. 3.20.3). As in last section, all analyses are made after 10 minutes of simulation. Increasing the albedo without changing the daytime, since the building surfaces reflect more solar radiation, results in a decrease in wall temperatures (comparison of Fig. 3.20.1 and Fig. 3.20.2). In the late afternoon (1600 LST), the sun has almost shifted around to the opposite position, therefore the shaded zones appear in front of the windward side and the heated walls are mostly the leeward sides. Figure 3.21 compares

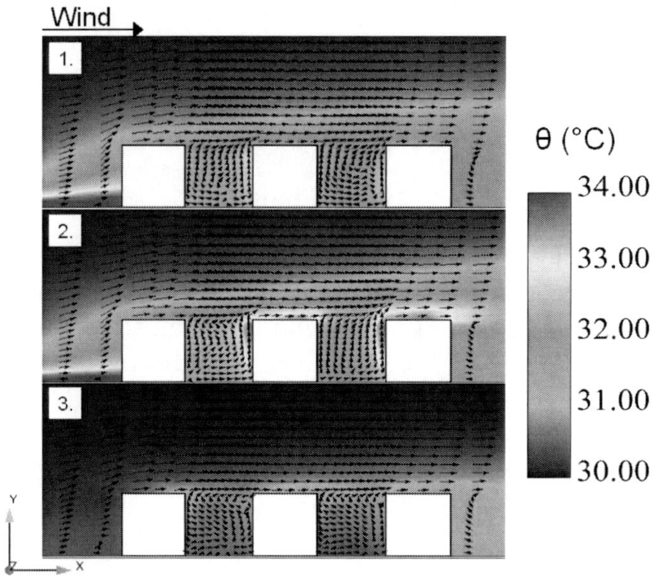

Figure 3.21. Comparison of the mean airflow potential temperature (θ) (°C) on the center-plane for three radiative transfer conditions: 1) a 0900 LST sun position with an albedo of 0.6 for the building walls; 2) same as 1) but with an albedo of 0.1 for the building walls; 3) same as 2) but with a 1600 LST sun position.

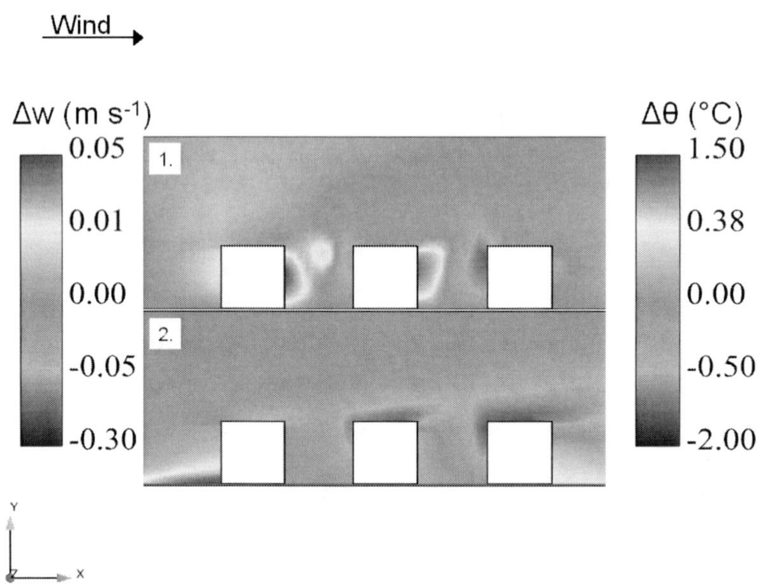

Figure 3.22. Difference with a 0900 LST sun position between 3D radiative transfer with an albedo of 0.6 and with an albedo of 0.1 for the building walls on the center-plane: 1) difference in vertical velocity (Δw) ($m\,s^{-1}$); 2) difference in potential temperature ($\Delta\theta$) (°C).

the distribution of the potential temperatures and the normalized velocity vector maps projected onto the center-plane under the three radiative transfer conditions listed above. In the morning, when increasing the albedo, the sunlit surfaces (windward walls and roofs) are less heated, and consequently the airflow is fairly less heated close to these areas (Fig. 3.21.1 and Fig. 3.21.2). To emphasize this further, the differences between the case with an increased albedo of the building walls and the reference one have been plotted: in vertical velocity (Fig. 3.22.1) and potential temperature (Fig. 3.22.2). The ground close to the first windward wall and the last leeward wall is more heated because it receives more reflected solar radiation from the building walls. As a result, the potential temperature is locally increased. Increasing the albedo of the walls not only locally changes the airflow pattern ($-0.05 \sim 0.30\ m\ s^{-1}$) but also the heating pattern: warmer ground and cooler windward walls and roofs.

In the late afternoon (1600 LST), in comparison to 0900 LST, the leeward walls are sunlit and are more heated than windward walls (Fig. 3.21.3). In comparison to the neutral case (Fig. 3.18.1), the airflow structure seems to be less disturbed than when the sun shines on the windward side and the potential temperature is not increased significantly.

Quantitative Analysis

In order to clarify and quantitatively analyze the thermal impact on the airflow, several vertical profiles of different variables are plotted at different positions in the domain. Figure 3.23a-f show the comparison of the vertical profiles of potential temperature, normalized vertical velocity and TKE, respectively, 0.24 m in front of the windward side and 0.28 m behind the leeward wall of the last building (from the left-hand side to the right-hand side) under different thermal models (i.e., neutral, imposed temperatures and different radiative transfer parameters). Taking into account the thermal stratification in the non-neutral cases, the airflow shows a large variability (Fig. 3.23a-f). Close to the windward wall, compared to the neutral condition and except in the radiative transfer case with a 1600 LST sun position, the potential temperature is largely increased near the ground level ($\sim +2.4°C$). From $z/H = 0.2$, the potential temperature rises and reaches its maximal value within the range $z/H = 0.8 \sim 0.9$. After that, the airflow cools down and the profiles tend to the neutral case at about $z/H = 3$. Near the last leeward wall, the potential temperature difference at the ground level between the radiative reference case and imposed temperatures case is about 2.5°C (Fig. 3.23d), which is due to the shading effects behind the last buildings not taken into account in the imposed temperatures case. When taking into account the thermal stratification, the vertical component of the velocity shows some variability (Fig. 3.23b and 3.23e). Without heating, the vertical velocity turns negative forming a downdraft flow in the canyons close to the windward sides or over the canyons close to the leeward sides. On the contrary, with thermal transfer, the vertical velocity is always positive and a strong updraft flow is formed from $z/H = 0.2$ up to the roof level. The impact of the heating on the TKE (Fig. 3.23c and 3.23f) is also important at these low wind speeds. The thermal gradients near the windward wall and near the leeward wall induce a difference in the TKE production.

Close to the windward wall, choosing the imposed surface temperature on the order of magnitude of the ones from the radiative transfer case makes this case agree well with the

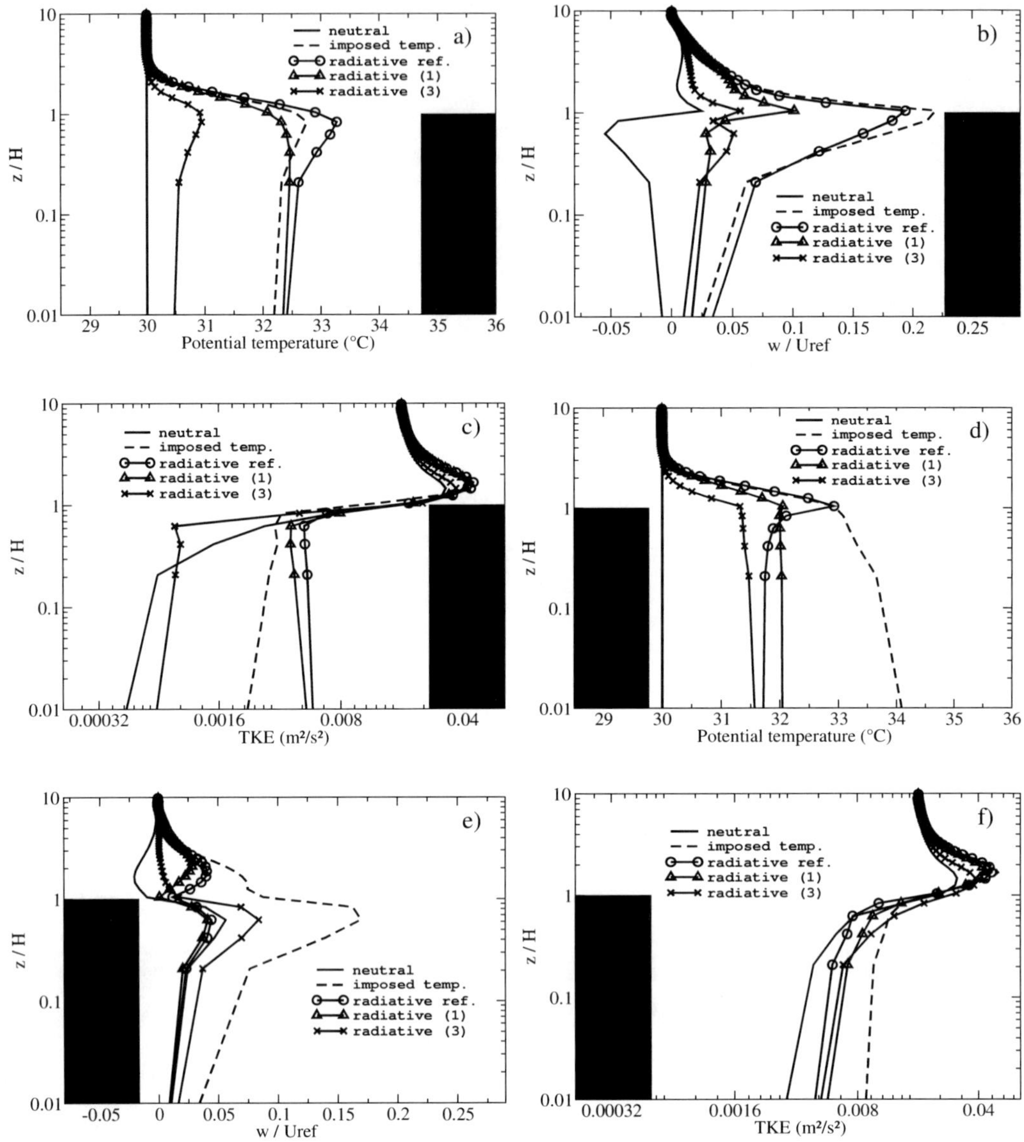

Figure 3.23. Vertical profiles of (a) potential temperature (θ) (°C); (b) normalized vertical velocity (w/U_{ref}); (c) turbulent kinetic energy (TKE) ($m^2\ s^{-2}$) 0.24 m in front of the last windward wall (from left-hand side to right-hand side) under different thermal conditions (full line: neutral; dashed line: imposed temperatures; circle symbol (i.e., radiative ref.): a 0900 LST sun position with an albedo of 0.1 for the building wall; triangle symbol (i.e., radiative (1)): same as radiative ref. but with an albedo of 0.6 for the building wall; cross symbol (i.e., radiative (3)): same as radiative ref. but with a 1600 LST sun position). (d), (e) and (f) same as (a), (b) and (c) but 0.28 m behind the last leeward wall.

radiative transfer case on the airflow properties. Significant differences are found close to the leeward side due to different treatments of the ground temperature at this area. These differences lead us to conclude on the importance of the non-uniform surface heating configuration with the 3D radiative model, linked to the full flow simulation.

For the radiative transfer cases, a late afternoon sunlit position (1600 LST) and using a higher albedo value for the walls both resulting in smaller heating. However, in comparison to the morning (0900 LST) with an albedo of 0.6, the late afternoon sun position with an albedo of 0.1 shows a weaker impact on the airflow (Fig.3.23a, 3.23b, 3.23c). Compared to the neutral case, the increase of the potential temperature is less than 1°C in the cavity (Fig. 3.23a). Moreover, the variability of the vertical velocity and the TKE are also moderate along the windward wall (Fig. 3.23b, 3.23c). Behind the last building, the heating of the leeward wall has less influence on the airflow pattern than the heating of the windward one.

4.4. Summary

In this section, the effects of different wall heating conditions on the airflow within an idealized canopy have been investigated in a low wind speed case, using new atmospheric radiative and thermal schemes implemented in the 3D CFD code (*Code_Saturne*). Firstly, the results show the importance of the thermal stratification effects in urban areas for low wind speed and especially the contribution of 3D radiative transfer within the canopy. The thermally induced motion is combined with the mechanically induced motion formed in the canyon, mainly resulting in increased upward motion. In this short-term study (10 minutes with a reference velocity of 1 $m\ s^{-1}$), the radiation heat transfer plays a major role in the surface energy balance (SEB). For instance, the net radiation flux (Q^*) on the windward wall of the center building is in the range of 240 $\sim$ 480 $W\ m^{-2}$ versus convection which is only about 20 $\sim$ 50 $W\ m^{-2}$. Because of the shading effects, the multi-reflections and the infrared emissions, 3D atmospheric radiative transfer induces a non-uniform and asymmetrical wall heating which, in comparison to the uniform and symmetrical wall heating from imposed surface temperatures conditions, modifies the airflow pattern. The maximum difference close to the leeward walls reaches 0.3 $m\ s^{-1}$ for the vertical velocity and 2 °C for the potential temperature.

Secondly, under radiative transfer conditions, the influence of the sun position and different values for the albedo of the building have been also discussed. The results show that changing the physical parameters of the walls have an important effect on the potential temperature and motion. Increasing the albedo from 0.1 to 0.6 (e.g., painting a dark surface to white) reduces the net radiation flux term in the SEB, as a result, the potential temperature close to the windward wall or roof may decrease by 2 °C. More examples, using different building materials (e.g., wood, brick) may modify the Q^* term (difference in the infrared emission) and the conductive heat flux (Q_{cond}) term (difference in the conductivity or thickness etc.), therefore, has a potentially important influence on the airflow pattern. By describing simulations with the same building physical parameters but different sun positions, solar heating of the leeward wall (at 1600 LST) has less influence on the airflow pattern than the heating of the windward wall (at 0900 LST) for this particular setup.

However, this discussion is based on an idealized urban area. In a real urban environment, the complex geometry has a strong influence on the wind field. Moreover, the simulation results were not compared to any experiment. Using a real measurement dataset may be appropriate to qualitatively evaluate the numerical simulations and enrich the discussion. In order to focus on topics related to microscale dispersion in irregular canopy and complex urban environments, in next Chapter an evaluation of this model on real urban areas will be presented.

Chapter 4

Application II: Real Urban Areas

In the previous chapter, the urban area is idealized. At microscales, small irregularities can break the repeated flow patterns found in a regular array of containers with identical shapes. In addition, uncertainties associated with the thickness and the properties of the material of the container wall limit validation of the results. That is the reason why the coupled thermal-radiative-dynamic model need to be evaluated on a district of a real urban in this chapter.

1. The CAPITOUL Campaign

Existing canopy models often use a statistical representation of buildings which is generally obtained through quantitative field survey or qualitative estimates. But in performing this geometric simplification there is no way to ensure that the simplified geometry locally match the actual city. Thus, the objective of this section is to represent the energy and momentum exchanges in an existing city as realistically as possible. The Canopy and Aerosol Particles Interactions in TOulouse Urban Layer (CAPITOUL) experimental dataset (Masson et al., 2008) has been chosen to validate the developed model as completely as possible with a large available experimental dataset in a real urban environment. Hereafter, I first give a brief overview of the experimental campaign. Then I present the mesh developed for the city center and the simulation conditions for the selected day of the campaign. Afterwards, I discuss the simulation results with the observations including surface brightness temperatures, infrared pictures taken by an airborne infrared thermal-graph, sensible heat flux, radiation flux and friction velocity data.

1.1. Overview of CAPITOUL Field Experiment

The CAPITOUL campaign is a joint experiment, organized by the Centre National de Recherches Meteorologiques and other partners (Laboratoire d'Aérologie, Laboratoire des Mécanismes et Transferts en Géologie and ORAMIP local air quality agency), which took place in Toulouse in the southwest of France (43°36'16.21"N, 1°26'38.5"E) from February 2004 to February 2005. It is an effort in urban climate, aiming to document the energetic exchanges between the surface and the atmosphere, the dynamics of the boundary layer over the city, and the interactions between the urban boundary layer and aerosol chemistry.

A general view of the experiment, describing the goals, experimental setup and a summary of the results are given by Masson et al. (2008).

Objectives and Description of the Site

The old downtown area with an area of approximately 3.5 km^2 is generally made up of buildings 4 to 5 stories high with the walls primarily composed of brick and stone, with roofs of clay tile (Fig. 4.1). The surrounding landscape is relatively flat with small rolling hills. Study of the energetic exchanges between the surface and the atmosphere was one of the objectives (Moscicki, 2007; Pigeon et al., 2008). I briefly describe the data used in this study.

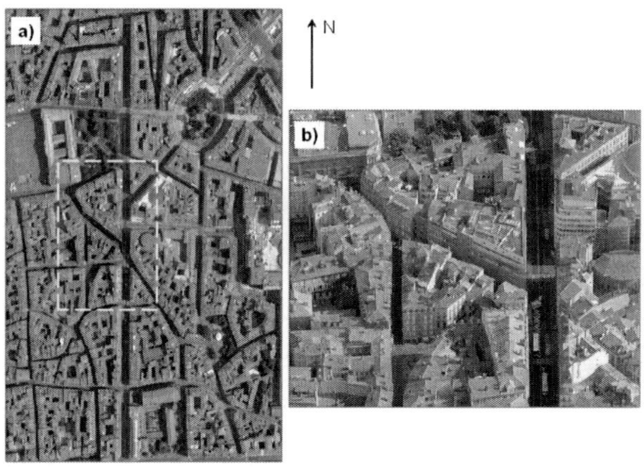

Figure 4.1. Aerial view of downtown Toulouse, France: a) main study area, from Google Maps; b) zoom in the selected area (in a contour), from Bing Maps. The north is indicated.

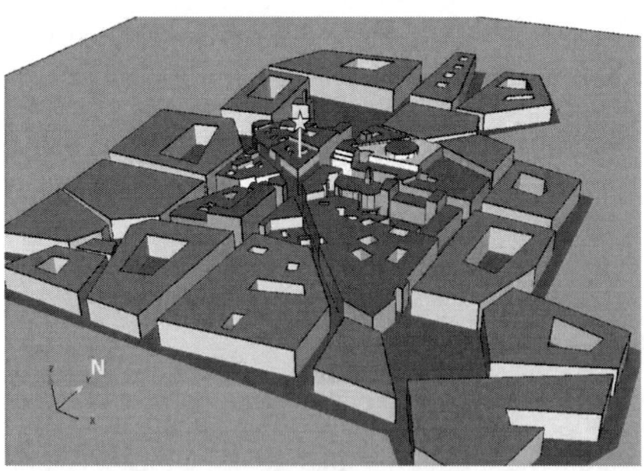

Figure 4.2. Central site area geometry processed by ICEM CFD with the meteorological mast position (star). The north is indicated.

Meteorological Data. The study area is mainly located in the central site of Toulouse around the corner of the two streets, Alsace-Lorraine and Pomme (contour in Fig. 4.1a and Fig. 4.1b). In this neighborhood, vegetation is very scarce and buildings are around 20 m high (Pigeon et al., 2008). The base of the meteorological mast was on a roof at a height of 20 m, with the top of the mast being 47.5 m above the road (position shown in Fig. 4.2). It provided data including short- and long-wave radiation flux, sensible heat, latent heat, air temperature, wind speed and direction etc. continuously from mid-February 2004 to early March 2005. All meteorological variables were sampled at one-second intervals and were recorded as one minute averages (Masson et al., 2008).

Infrared Surface Temperature Measurements and Aircraft Data. A total of ten InfraRed Thermometers (IRTs) were affixed to balconies or booms to record the surface temperatures of the roads, walls, and roofs of the canyons. Described in details in Moscicki (2007), four IRTs were positioned in the Alsace canyon to record surface temperatures of the two walls, the road, and a roof. Three IRTs were located in both the Pomme and Remusat canyons to observe the temperature of the two walls and the road in each canyon.

InfraRed Thermal (IRT) airborne images were also obtained during several Intensive Observation Periods (IOP) with 2 airborne cameras on board of a Piper Aztec PA23 aircraft over the study area (flight 430: 0749 - 0816 UT; flight 431: 1115 - 1150 UT; flight 432: 1348 - 1423 UT). The speed of the aircraft was 70 $m\ s^{-1}$ and the camera acquisition frequency was 4.3 frames per second. The flight height was about 460 m, which results in a resolution between 1.5 and 3 m depending on the sight angle (Lagouarde et al., 2010; Hénon et al., 2012).

1.2. Simulation Setup

Concerning urban canopy energetics issue, several papers on the CAPITOUL project have been published. Pigeon et al. (2007, 2008) present the modeling of the anthropogenic heat flux by the TEB urban scheme (Masson, 2000) and its validation against anthropogenic fluxes estimated by a new method using standard surface energy balance measurements. Based on the use of SOLENE model coupled with a 3D model of the city providing the information about the actual structure of the urban canopy, Lagouarde et al. (2010) simulate the IRT anisotropy and the directional surface temperatures over the Toulouse city. Gastellu-Etchegorry (2008) imports the Toulouse urban database as a DART scene (Gastellu-Etchegorry et al., 2004) and tests the DART-TEB model for simulating remote sensing images and the radiation budget of urban canopies. Using SOLENE software, Hénon et al. (2012) assessed the case of a small district of the city center for four independent sets of measurements for two complete diurnal cycles, in summer and in winter by making comparisons with the thermal-radiative simulations.

Whereas most earlier works estimate the sensible heat flux with simplified convection models, the heat transfer coefficient h_f is usually considered as a constant or a simple function of height. In order to model the microscale heat transfer with more accuracy by determining the surface-air thermal gradient that controls convective heat transfer and also to examine the mechanisms (e.g., complex topographic influences on air motions), the

CAPITOUL simulation has been performed with *Code_Saturne* to investigate the thermal-dynamic impacts on the local atmosphere (Qu et al., 2014, 2013).

Choice of the Computational Domain and Mesh Strategy

From the CAPITOUL experiment, the day of July 15th, 2004 has been selected. Considering Alsace-Lorraine and Pomme roads as the center of interest in the computational domain (Fig.4.1), the dimension of the 3D simulation domain is $891 \times 963 \times 200$ m. The information about the 3D structure of the urban canopy was provided by the administrative authorities of Toulouse who made the 3D database of the city available for the CAPITOUL project.

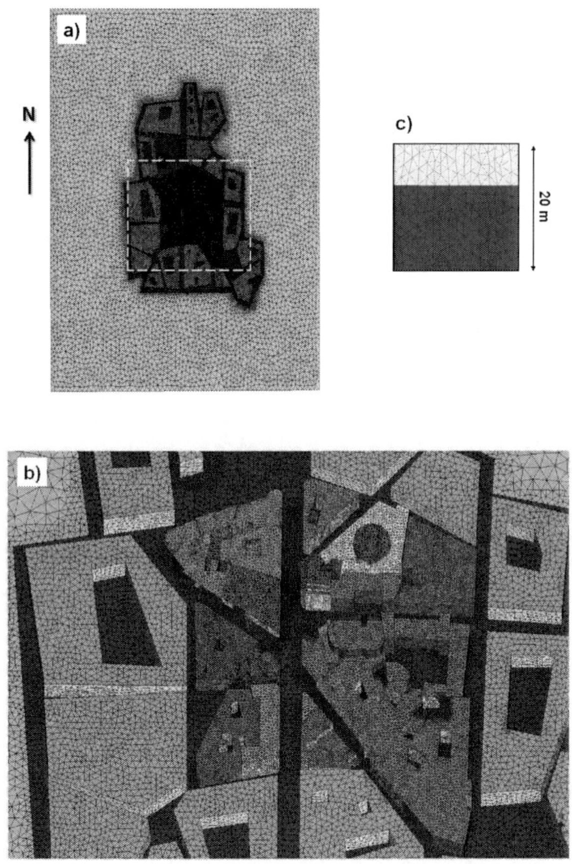

Figure 4.3. Tetrahedral mesh on the central area: a) whole area; b) zoom in the selected area a) (contour); c) zoom in a Monoprix building wall. The north is indicated.

First, the urban database (AutoCAD format) of the Toulouse town hall was imported into the commercial mesher ICEM CFD. The urban elements in the database are not individual houses or buildings but a group of the walls and roofs including a large number of internal fine walls which are unnecessary to be meshed. Moreover, the urban elements have a variety of heights but no soil element. So before the meshing step, it is necessary to do a preparatory work on the geometry. After a series of geometric optimization (remove the internal surface, simplify a part of details, create the ground then project the buildings onto it ...), a proper geometry topology is built as shown in Figure 4.2. In real urban environment, all obstacles cannot be resolved with sufficient detail, but their impact needs to be parametrized. The strategy for this study is described as follows. From the boundary of the domain to the center, more geometric details are progressively retained. That is, the buildings at these Alsace-Lorraine and Pomme streets in the center study area are modeled with fine details. Then, the surrounding buildings next to the center study area are simplified as urban blocks. Finally, the buildings in the region outside are treated with a high roughness value in this study. The volumetric mesh used here is an unstructured grid of about 4.5 million tetrahedral cells. The grid resolution varies from 0.5 m near the center to 10 m far from the center zone (Fig.4.3).

Initial and Boundary Conditions

The wind inlet boundary conditions are determined from measurements, using a meteorological mast (Fig. 4.2) which gives the wind velocity every 2 hours (Fig. 4.4a), wind direction (Fig. 4.4b), and potential temperature profiles (Fig. 4.4c).

From some Toulouse pictures, the roughness value is estimated depending on its location and set value for both dynamic roughness length and thermal roughness length. Based on Pigeon et al. (2008), the thermal properties are set such as surface conductivity and thickness. The model parameters are summarized in Table 4.1. Since the values from Pigeon et al. (2008) are averages over the 500-m radius around the surface energy balance station, watching some Toulouse pictures from Google Maps, the author furthermore classifies four wall painting color (rose, gray, whitewash and white) for the buildings in the center area (Fig. 4.2) to estimate the albedo. Their values are given in Table 4.2.

1.3. Results and Discussion

The whole 24-hour simulations took a total of 15 hours with 64 processors.

Comparison of IRT Pictures

An IRT picture from for the aircraft flight 432 of July 15th, 2004 at 1412 UT is shown in Figure 4.5a. Figure 4.5b and 4.5c depict respectively the modeled brightness temperature and surface temperature with the thermal-radiative-convective full coupling. It should be noted that it is difficult to compare value by value because of the simplification of the geometry. However, since different albedo are distinguished from painting colors, the model reproduced well the heterogeneity of the brightness temperature, especially at roof level (Fig. 4.5b). It should also be noted that if more details were kept, especially regarding some slopes on the central roof in the 3D model to create a more detailed distribution of the shadows on the roof, the simulated brightness roof temperature would be closer to the observations. The value of the modeled surface temperatures (Fig. 4.5c) is obviously larger than the modeled brightness one (e.g., about 3 °C difference on the roof), because from equation 2.35, we can see that the brightness temperature is almost proportional to the product of the surface temperature and emissivity ($T_{br} \approx T_{sfc}\varepsilon^{1/4}$, the emissivity is smaller than 1).

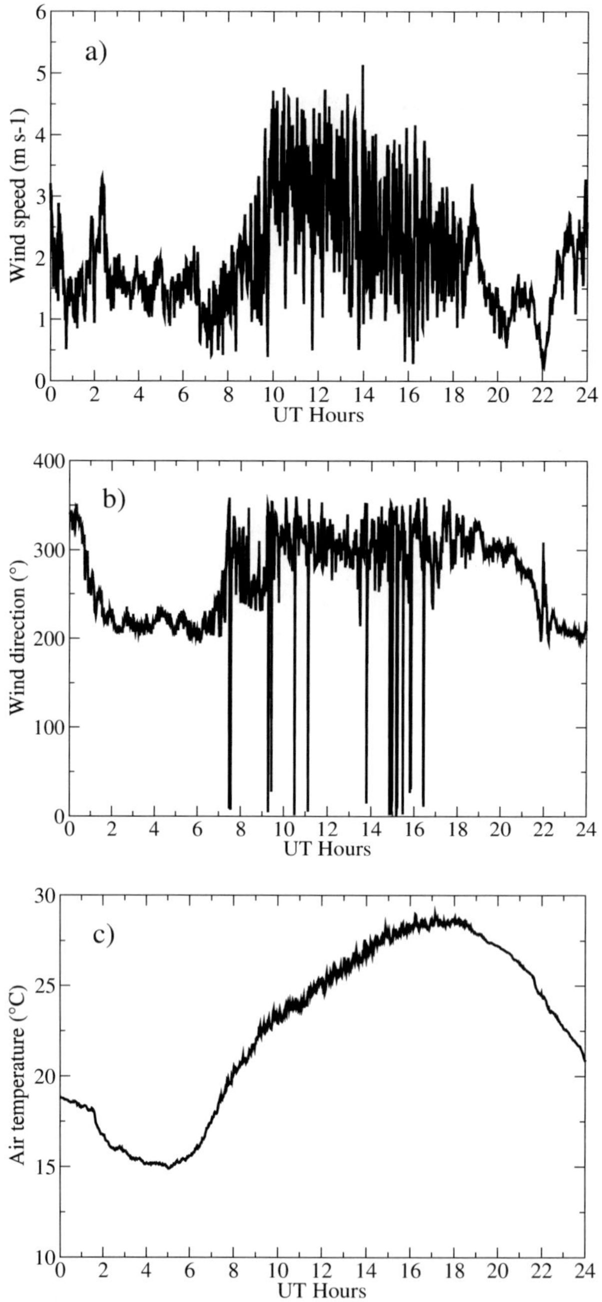

Figure 4.4. Meteorological data at 47.5 m above the road from the CAPITOUL experiment for July 15th 2004: a) Wind speed ($m\ s^{-1}$), b) Meteorological wind direction (°), and c) Air temperature (°C).

Figures 4.6 from (a) to (d) show the measured and modeled brightness temperatures for the aircraft flight 431 at 1138 UT, compared with three modeling approaches: thermal-radiative model only (Fig. 4.6b), thermal-radiative and a constant heat transfer coefficient (Fig. 4.6c),

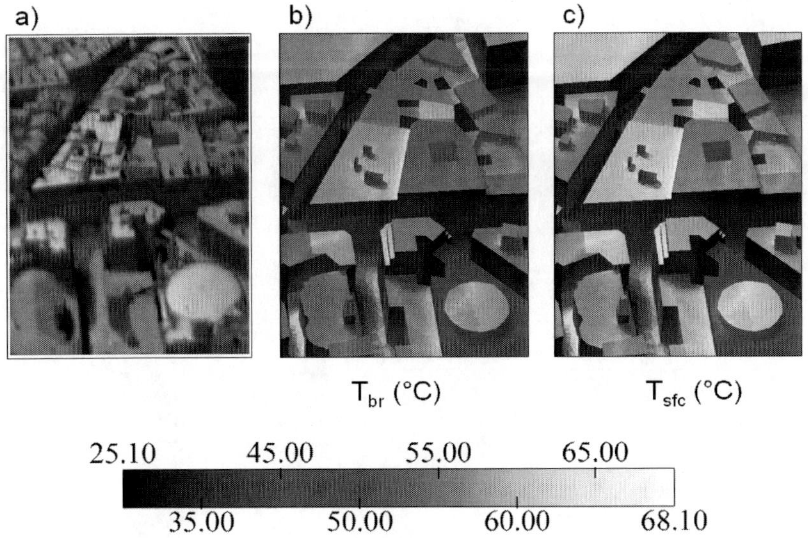

Figure 4.5. Comparison between the Infrared Thermal (IRT) airborne images, simulated brightness temperatures and surface temperatures of July 15th 2004, at 1412 UT during flight 432 (Lagouarde et al., 2010): a) IRT picture (189 × 118 pixels), source from Hénon (2008); b) modeled brightness temperature with full thermal-radiative-dynamic coupling; c) same as b) but for modeled surface temperature.

and thermal-radiative-convective full coupling (Fig. 4.6d) using the hybrid thermal scheme for the streets and walls. The constant heat transfer coefficient (Fig. 4.6c) is set similar to Hénon et al. (2012). The result of the first simulation (Fig. 4.6b, thermal-radiative scheme alone) shows that the brightness temperatures are obviously higher than measurements. For a visibility reason, an additional black contour is drawn to show the building structures. Especially on the roof, the difference is more than 25 °C. The modeled T_{br} is out of color range.

Either taking a constant heat transfer coefficient, i.e., assuming a constant wind field in the domain or taking a variable heat transfer coefficient with full coupling i.e., more realistically modeling the wind field, as expected, the results (Fig. 4.6c and 4.6d) show a much better agreement with observation (Fig. 4.8a) in comparison to thermal-radiative model only case (Fig. 4.6b). From this IRT picture with this view (Fig. 4.6a), in spite of the fact that not all the elementary details were kept on the roofs with different orientations and slopes, the simulated temperatures represent well the spatial variability of the temperatures.

For the variable h_f case, the model-observation difference rarely exceeds 10 °C (Fig. 4.6d). It should also be noted that in the measurement (Fig. 4.6a) for the same roof and same orientation, the temperature may differ by more than 5 °C, for instance at the bottom right of the image. This may be due to heterogeneities in materials and geometry which can not all be accounted for by modeling each individual area (or even every detail). For building walls, either shaded or sunlit, the difference between measurement (Fig 4.6a) and simulation (Fig. 4.6d) is generally less than 5°C. In the measurement, some horizontal

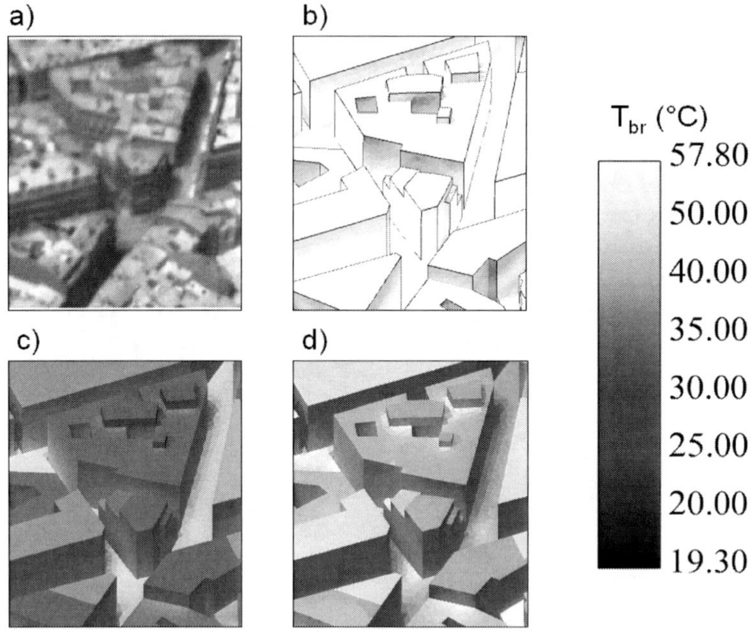

Figure 4.6. Comparison between the simulated brightness temperatures and Infrared Thermal (IRT) airborne images of July 15th 2004, at 1138 UT during flight 431 (Lagouarde et al., 2010): a) IRT picture (189 × 118 pixels), source from Hénon (2008); b) Modeled brightness temperature without taking into account the convection; c) Same as b) but with a constant heat transfer coefficient; d) Same as b) but with full thermal-radiative-dynamic coupling.

faces (e.g., buildings at left in the center) are relatively warmer than others. This may be due to some external structures (e.g., balcony) that are not modeled but were exposed to the sun and, therefore, received more solar heating (Hénon, 2008). Regarding the streets, a minimum of 3 cells were set for the width; the model is able to simulate the sharpness of the shadow (Fig.4.6c and 4.6d). The portion of the street brightness temperatures near the buildings is well reproduced. The averaged difference is less than 3 °C (Fig.4.6d). The simulated sunlit portion is underestimated by about 5 °C (Fig.4.6d).

The difference between constant h_f (Fig. 4.6c) case and variable h_f case (Fig. 4.6d) is also evident. These two different h_f are shown in Figure 4.7. With the chosen value (Fig. 4.7a), the constant h_f case shows an overcooling about 15°C at roof level, 4 to 5 °C at walls, 4 °C in shaded streets and 7 °C in sunlit streets (Fig. 4.7c). The difference is due to the overestimation of constant chosen value ($h_f = 12$) at this moment. Actually, the contribution of the variable h_f (Fig. 4.7b) may be more important on the surface and air temperature when modeling the temporal evolution. For this reason, the simulation is performed with full coupling for a diurnal evolution in the next section.

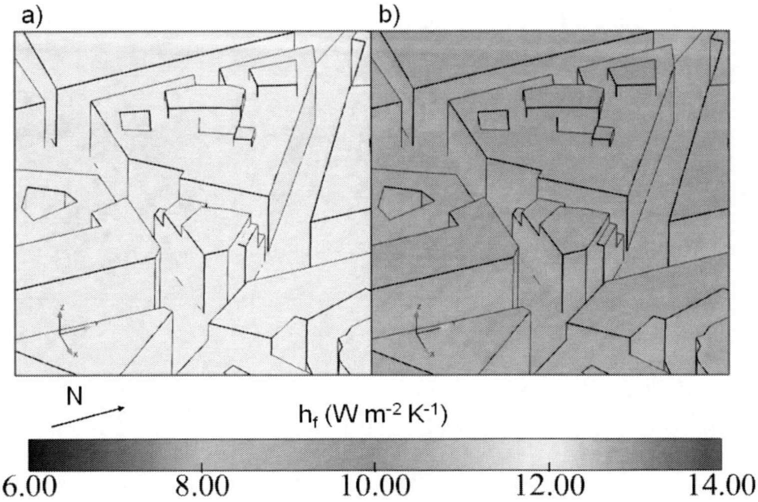

Figure 4.7. Comparison between two convective models with visualisation of heat transfer coefficient $h_f(W\ m^{-2}\ K^{-1})$ during same flight as Fig. 4.6: a) constant $h_f = 12$ corresponding to Fig. 4.6c; b) variable h_f corresponding to Fig. 4.6d. The north is indicated.

Comparison of the Local Diurnal Evolution of Brightness Surface Temperature T_{br}

Figures 4.8a to c present the model-observation comparison for the brightness surface temperatures of the diurnal cycle of July 15th, 2004. The simulation is performed with full thermal-radiative-dynamic coupling. Hereafter, unless specified otherwise, all the results refer to full coupling simulations. The infrared radiometers provided the measured brightness temperatures. Their fixed positions around the central building of the study district are shown in Figure 4.9. Overall, the diurnal evolutions of the brightness surface temperatures at the local positions in the scene are correctly simulated. For the faces of the Alsace street (Fig. 4.8a), an over-cooling about 5 °C appears during the evening (1800 to 2400 UT). However, for the Alsace west face, the model predicts a higher brightness temperature (Fig. 4.8a) from 0600 to 1200 UT. By using SOLENE model, Hénon et al. (2012) also report a similar difference in surface temperature. They explain that this may be due to a sensor underestimation. The ground temperature is computed by the force-restore method which is well adapted for the soil model. The bias on the Alsace road (Fig. 4.8a) can be explained by the approximation of the modeled shadow. Indeed, values taken from the selected cell might be quite different from its neighborhood values.

The best-simulated brightness temperature is the Pomme route one (Fig. 4.8b). The primary model-observation disagreement occurs on Pomme northeast face (Fig. 4.8b) during the afternoon. The bias reaches 8 °C. This can be explained by the same way for the Alsace road. Moreover, from some photos, there are numerous windows with white blind on this side of the wall. The infrared radiometer might detect a position where white blinds were closed, therefore more solar radiation flux was reflected. Or windows were opened, the ventilation in the canopy led to a lower internal building temperature which decreased

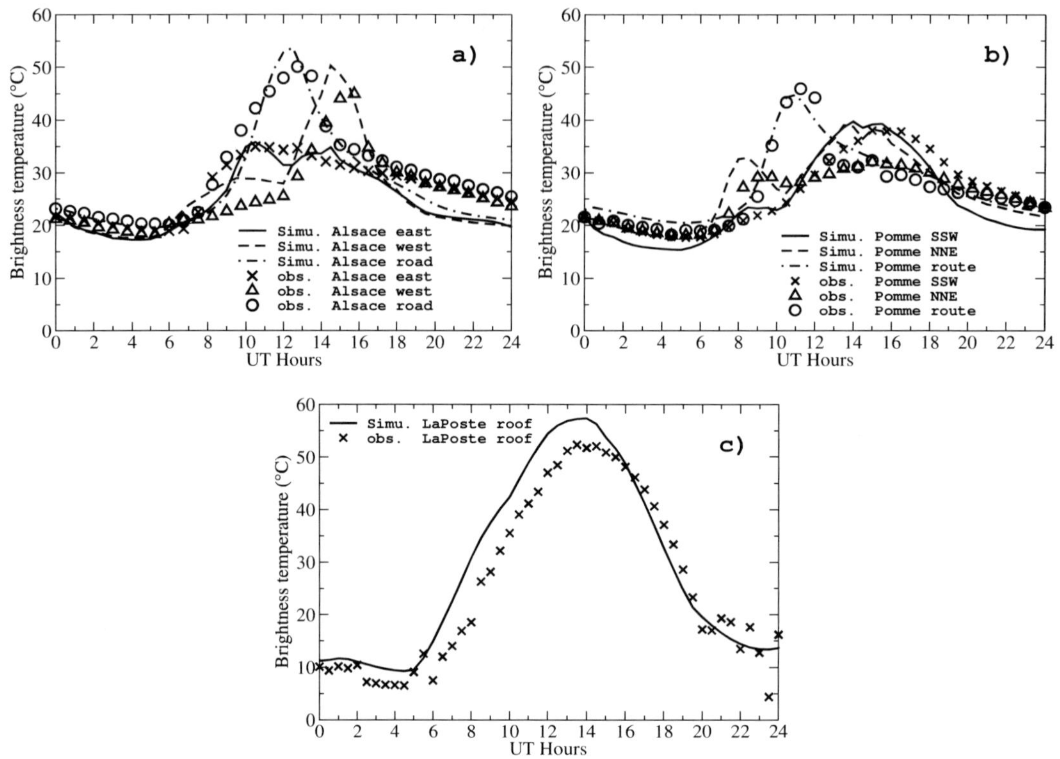

Figure 4.8. Evolution of brightness surface temperature of different positions of the infrared thermometers during a diurnal cycle (symbols: Measurements; lines: Simulations).

the external temperature. Finding a solution is a hard task. Firstly, there is an uncertainty in the observation. Secondly, this wall needs to be modeled in more detail, maybe including the in-front walls and windows.

The modeled roof brightness temperature displays a good agreement with observation before sunrise and from late afternoon to midnight (Fig. 4.8c). However, the model exhibits a significant advance in warming when the roof starts being sunlit. Actually, the sensor detecting roof brightness temperature was fixed at a northern building's roof from the mast (Fig. 4.9c and Fig. 4.10). This building is treated as a simple block in the 3D model (Fig. 4.2). However, from Figure 4.9c and 4.10, the complex structure of the roof can be observed. In particular, there are some small obstacles which are higher than the measured roof to the east but are not represented in the simulated scene. During the sunrise, the roof might be shaded due to these non-modeled detail structures. From the late afternoon, the sun shifts to another side. Since no more small obstacles are higher than the measured roof, the bias disappears.

Outgoing Long-Wave Radiative Flux $L^{\uparrow}$

Figure 4.11 shows the model-observation comparison of long-wave radiation flux. The difference between the modeled outgoing long-wave radiation flux at roof surface (full line) and on the mast (dashed line) is generally less than 20 $W\ m^{-2}$. However, both of them give

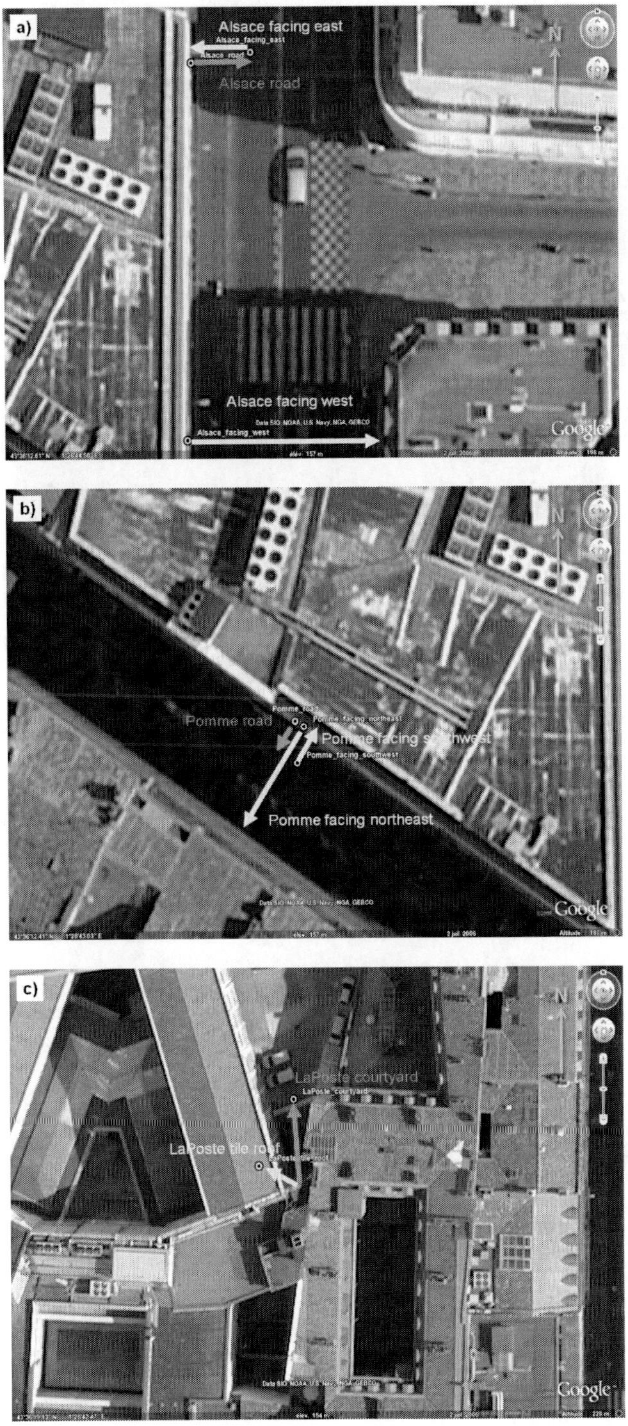

Figure 4.9. From a) to c), different positions of the infrared thermometers. The north is indicated.

an underestimation of about 20 $W\ m^{-2}$ during the night and a maximum overestimation of about 100 $W\ m^{-2}$ in the daytime. In fact, these differences are related to the error on the

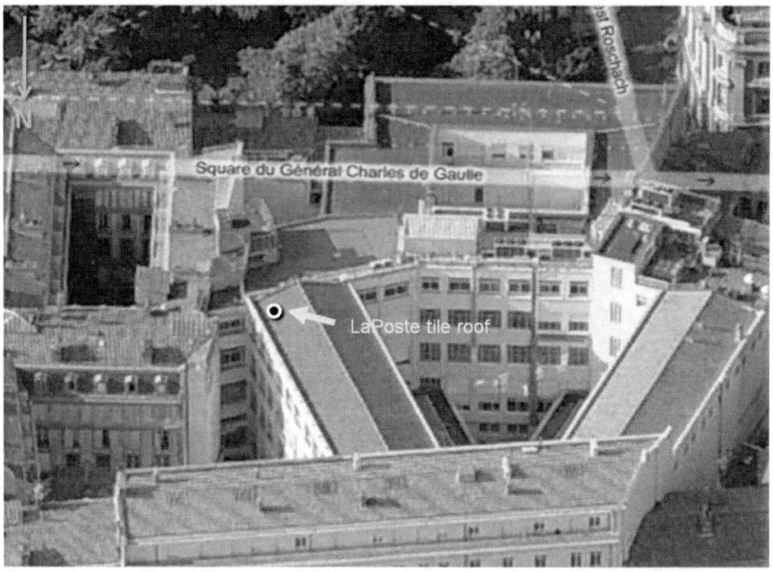

Figure 4.10. Aerial view of IRT position on the La Poste roof from Bing Maps. The north is indicated.

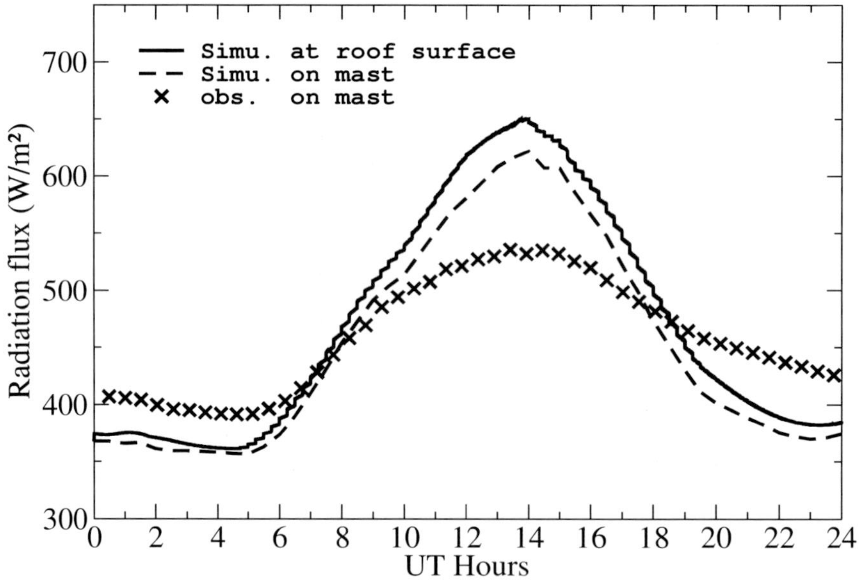

Figure 4.11. Diurnal variation of outgoing infrared flux calculated at roof surface (full line), at the mast level (dashed line) and measured on the mast (cross symbol).

roof temperature, i.e., underestimation in the night and overestimation in the day. It should be recalled here, from Eq. 1.6 the outgoing infrared radiation flux highly depends on the surface temperature because of the term of $\varepsilon\sigma T_{sfc}^4$.

To complete the discussion, the model offers the advantage of being able to visualize the radiative flux in the computational domain. Fig. 4.12 illustrates the distribution of the

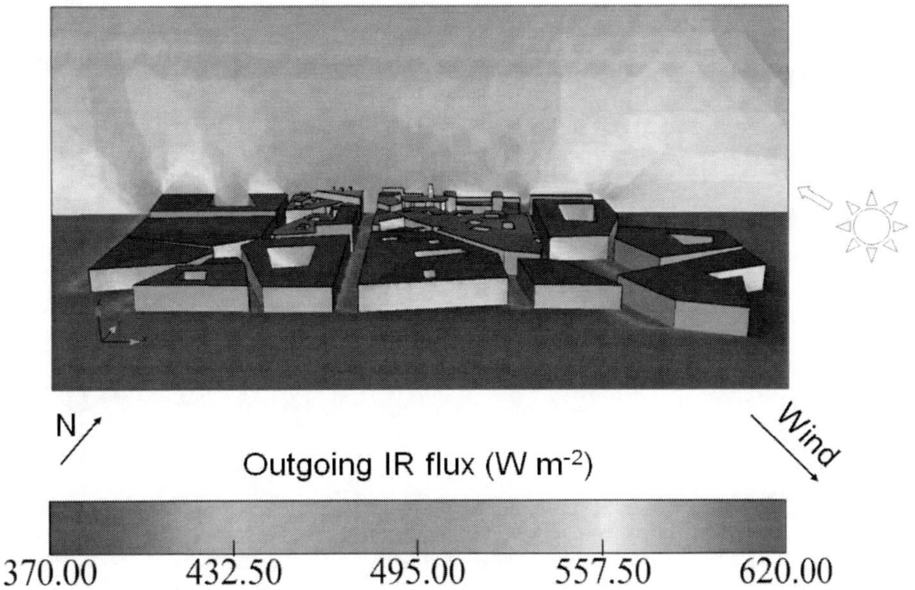

Figure 4.12. Visualization of the outgoing infrared flux on the center-plane and surfaces at 1030 UT. The north, wind and sun direction are indicated.

outgoing infrared flux at 1030 UT on the vertical center-plane and on the building surfaces. Significant variability can be observed on the center-plane. Relatively high values are found at the horizontal solid-air interface due to the fact that horizontal surfaces are warmer than the vertical surfaces in the daytime.

Outgoing Global Radiative Flux $S^{\uparrow}$

Figure 4.13 compares the modeled upward global solar flux with the measured one. The non-zero measured nighttime solar flux may be due to the sensor errors or the artificial lights from shop-windows, cars and street-lamps (Hénon et al., 2012). The upward solar flux can be estimated by

$$S^{\uparrow} = \alpha(S_D + S_f + S_e) \tag{4.1}$$

where S_D ($W\ m^{-2}$) is the direct solar flux, S_f ($W\ m^{-2}$) the solar flux diffused by the atmosphere above, S_e ($W\ m^{-2}$) the flux diffused by the environment (i.e., from multi-reflection on the surfaces).

For this gray roof (Fig. 4.9b), an albedo of 0.15 was set (Table 4.2). The same value is also proposed by Pigeon et al. (2008) and is used in Hénon et al. (2012). Compared to the observation, in spite of about 20 $W\ m^{-2}$ higher values at noon estimated by the model (full line), the agreement between measurement and modeled results of the outgoing solar flux on the mast (dashed line) is very satisfactory.

In the same manner as for the long-wave radiative flux, Figure 4.14 displays the distribution of the outgoing global solar flux at 1030 UT on the center-plane and on the surfaces. By visualization the propagation of the radiative flux in the fluid domain, we can better

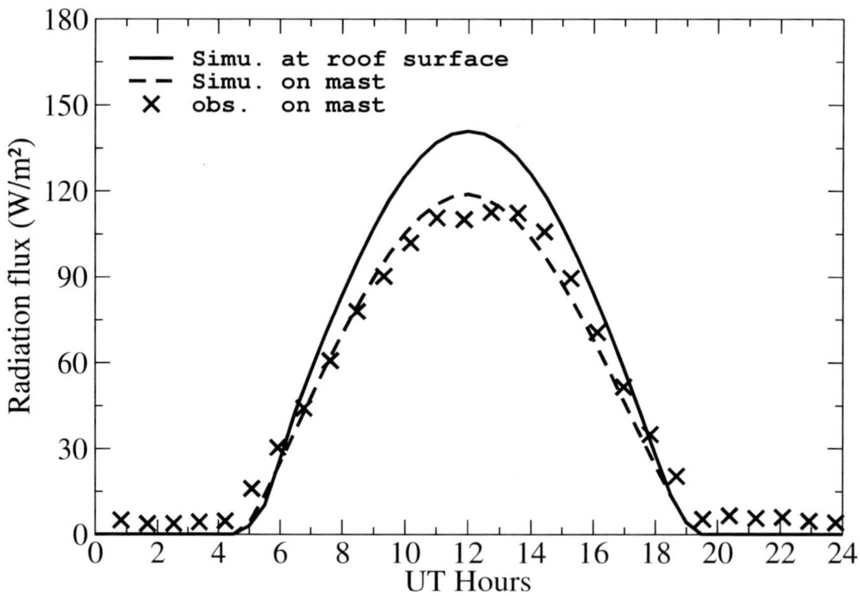

Figure 4.13. Same as Figure 4.11 but for outgoing solar flux.

understand the distribution of the shadow projected by the different structures. We can also readily identify surfaces with higher albedo since they reflect more solar radiation flux.

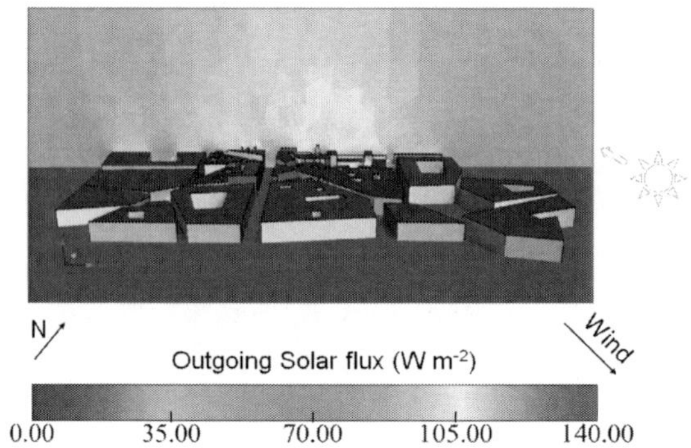

Figure 4.14. Same as Figure 4.12 but for outgoing solar flux.

Sensible Heat Flux Q_H

Surface sensible heat flux is the energy exchanged between a surface and the air in the presence of a surface-air thermal gradient. Modeling the sensible heat flux helps to determine both stratification effects on turbulent transport, and to estimate the surface temperature. At roof surface, the sensible heat flux Q_H can be parameterized as:

$$Q_H = h_f(T_a - T_{sfc}), \tag{4.2}$$

in which h_f ($W\ m^{-2}K^{-1}$) is the heat transfer coefficient and T_a (K) the air temperature.

The sensible heat flux at the surface can be recorded easily in the simulation and be compared to the value measured on the mast computed using the eddy-covariance technique. It can be estimated at the mast level as follows:

$$\overline{w'\theta'} = K_t \frac{\partial \theta}{\partial z}, \tag{4.3}$$

where $\overline{w'\theta'}$ is the potential temperature fluctuation, and eddy diffusivity K_t ($m^2\ s^{-1}$) by using turbulent viscosity ν_t ($m^2\ s^{-1}$) and constant turbulent Prandtl number Pr_t as $K_t = \nu_t / Pr_t$. In this work, Pr_t is set to 1.0. Hence, $Q_H = \rho C_p \overline{w'\theta'}$.

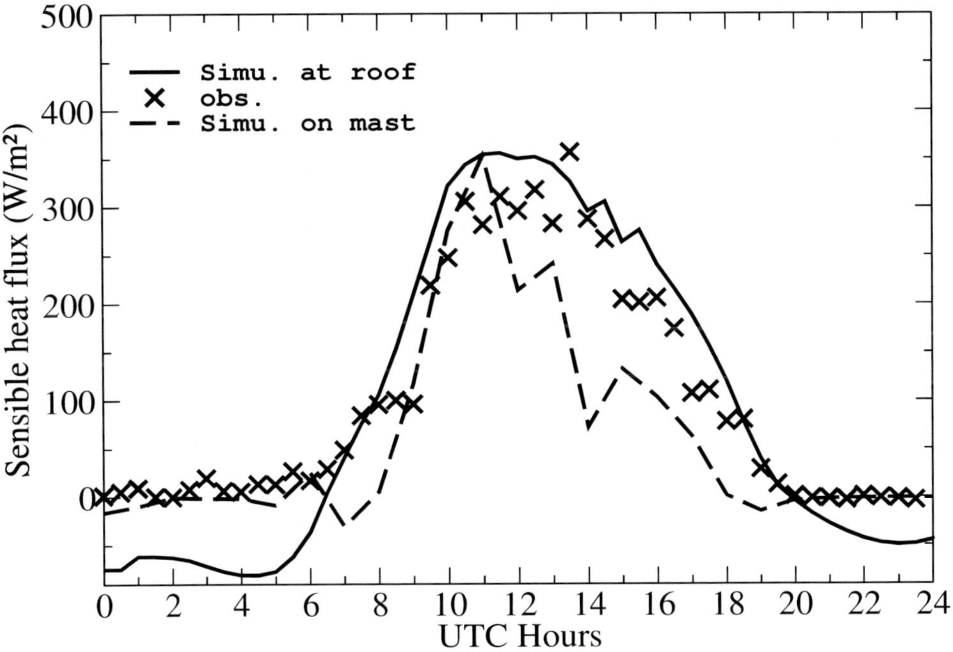

Figure 4.15. Same as Figure 4.11 but for sensible heat flux.

Figure 4.15 depicts the comparison between the time evolution of calculated sensible flux and on the mast observation. At roof, differences between observed and calculated values are less than 40 $W\ m^{-2}$, except for the night ($45 - 70\ W\ m^{-2}$). From midnight to 0700 UT, the wind is calm but does not have a zero speed (Fig. 4.4a) and the air temperature varies from 15 to 19°C (Fig. 4.4c). The observed sensible flux is very small (cross symbol in Fig. 4.15), this may be due to the fact that the air layer just above the roof has a temperature very close the roof temperature. However, the initialized roof temperature at this position is set to the same value as the measured roof temperature (11.13°C in Table 4.1). With a non-zero heat transfer coefficient and a higher air temperature, consequently, negative sensible heat flux is obtained at roof surface (full line in Fig. 4.15). For the rest of the day the comparison is rather good. At the mast location, reasonable agreement generally exists between calculated and measured values. The differences are more pronounced from 1300 UT to 1500 UT. The reason may be due to uncertainty from the measurements and not detailed geometries on the roofs, which has a local influence on the convective flux.

Model-Observation Comparison of Friction Velocity u^*

Friction velocity u^* ($m\ s^{-1}$) is also measured on the mast during the CAPITOUL experiment. The u^* at the roof surface is defined by the relation:

$$u^* = (|\tau_w/\rho|)^{1/2}, \tag{4.4}$$

where τ_w is the Reynolds stress at the wall ($N\ m^{-2}$), ρ ($kg\ m^{-3}$) the fluid density at the wall.

In order to compare with the observation on the mast, the kinematic momentum fluxes is used in the x and y directions ($\overline{u'w'}$, $\overline{v'w'}$), therefore the friction velocity can be evaluated as:

$$u^* = (\overline{u'w'}^2 + \overline{v'w'}^2)^{1/4}, \tag{4.5}$$

where $\overline{u'w'}$ and $\overline{v'w'}$ are given by:

$$\overline{u'w'} = \nu_t \frac{\partial \bar{u}}{\partial z}, \ \overline{v'w'} = \nu_t \frac{\partial \bar{v}}{\partial z}, \tag{4.6}$$

and ν_t ($m^2\ s^{-1}$) is the turbulent viscosity given by the $k - \varepsilon$ closure.

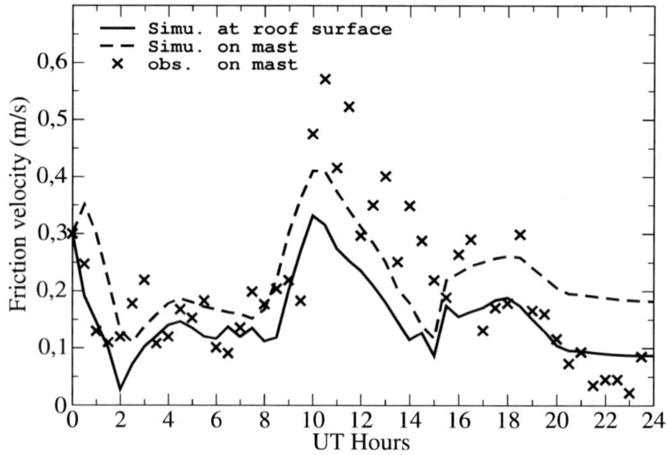

Figure 4.16. Same as Figure 4.11 but for friction velocity.

From Figure 4.16 it can be seen that the simulated friction velocities are reasonable in terms of magnitude. Modeled friction velocity at roof surface (full line) shows a good agreement with observation when the u^* is less than 0.2 $m\ s^{-1}$. Simulated u^* on mast (dashed line) can produce higher value (0.4 $m\ s^{-1}$). It seems that the model overestimates the nighttime (1800 – 2400 UT) u^* on the mast. However, estimates of u^* from the turbulence measurements by the sonic are known to be rather uncertain, especially when u^* is small. The difference between the modeled u^* at roof surface (full line) and at mast level (dashed line) is due to different methods of the estimation, and the numerical errors in the computed gradients.

1.4. Summary

In this section, the energy exchanges in a real city with the atmosphere have been investigated during the CAPITOUL campaign, using new atmospheric radiative and thermal schemes implemented in *Code_Saturne*. A preprocessing is realized including the optimization of the complex geometry and creation of a high-quality tetrahedral mesh for this study. It also requires determining the complex thermal parameters which take into account the actual variability of materials in the district. Based on the literature data, the building surfaces have been separated into 4 categories of albedo depending on painting colors.

First, simulations are evaluated with the comparison with Thermal InfraRed airborne images from two flights in the day of July 15th, 2004 during the CAPITOUL project. The result shows the importance of taking into account heterogeneities in materials and geometry to represent the spatial variability of the temperatures in complex urban areas. Then, the coupled thermal-radiative-dynamic model has been evaluated with CAPITOUL experiment including the comparison with the measured brightness temperature, sensible flux, radiative fluxes diurnal variation. Overall the agreement between measurements and model simulations are fair but can be improved in the future with more information. For the sensible flux, the model-observation nighttime bias is probably linked to the uncertainty of the estimation on the roof nighttime temperature. Similar explanation can be used for the comparison of the outgoing infrared flux, because results are sensitive to the surface temperature. In fact, due to the complex geometries, a comparison of the IR flux in a simple case might be appropriate to understand the difference and evaluating the model.

Better agreement is obtained for the comparison of the outgoing solar flux at mast level. For the modeled friction velocities, the difficulty is to capture the extreme values (highest or lowest). Small structures may have important influence on the computation of local brightness surface temperature, sensible flux, outgoing short- and long-wave radiation. The simulation results show the importance of modeling in details while doing local model-observation comparison.

2. The EUREQUA Campaign

This section describes another example of numerical CFD urban simulations, performed at the neighborhood scale in the framework of the EUREQUA project (Haoues-Jouve et al., 2015) for a full diurnal cycle including daytime solar heating and nighttime infrared cooling. Accomplishing the objective requires assembling a set of computational tools to take a real city's urban topography and construct a SEB simulation. The numerical mesh is constructed semi-automatically on the basis of the building database available for the entire country. Results are compared to thermal images and in situ measurements obtained during the three days first field campaign.

2.1. Overview of EUREQUA Project

The EUREQUA (Evaluation mUltidisciplinaire et Requalification Environnementale des Quartiers) is a French national research project that began in March 2012 with the aim of examining issues of the environmental redevelopment of the city at the local scale, in Mar-

Table 4.1. Parameters applied in the CAPITOUL simulations

Parameter	Unit	Value
Dynamic		
Timestep	s	10
Z_{ref}	m	47.5
U_{ref}	$m\ s^{-1}$	3.21
ϕ	$^{\circ}$	340
Roof z_0	cm	10
Wall z_0	cm	10
Street z_0	cm	2
Radiative		
Timestep	min	15
Roof z_{0_T}	cm	10
Wall z_{0_T}	cm	10
Street z_{0_T}	cm	2
α_S		0.08
ε_R		0.9
ε_W		0.92
ε_S		0.95
λ_R	$W\ K^{-1}\ m^{-1}$	0.83
λ_W	$W\ K^{-1}\ m^{-1}$	1.15
λ_S	$W\ K^{-1}\ m^{-1}$	0.69
e_R	m	0.09
e_W	m	0.3
e_S	m	1.25
μ_R	$J\ m^{-2}\ s^{-0.5}\ K^{-1}$	1006
μ_W	$J\ m^{-2}\ s^{-0.5}\ K^{-1}$	1348
μ_S	$J\ m^{-2}\ s^{-0.5}\ K^{-1}$	1330
Initial temperature values		
T_{air}	°C	18.86
Alsace East (outside/inside)	°C	22.36/22.94
Alsace Road (outside/inside)	°C	23.18/25
Alsace West (outside/inside)	°C	21.32/22.23
Pomme SSWest(outside/inside)	°C	24.70/25.55
Pomme Route (outside/inside)	°C	23.33/25
Pomme NNEast (outside/inside)	°C	29.06/29.78
La Poste Roof (outside/inside)	°C	11.13/10.36

seille, Paris and Toulouse in France, in areas where city remodeling projects are planned.

Table 4.2. Albedo classification of the buildings surfaces

Paint color	albedo
White	0.6
Whitewash	0.5
Rose	0.3
Gray	0.15

This project adopts an original methodological approach relying on a multidisciplinary team involving geographers, sociologists, atmospheric physicists, acousticians, and architects, in collaboration with officials of urban living.

The first phase of the project led to choosing three relevant districts in Toulouse, Paris and Marseille, namely those with specific environmental nuisances (microclimate, noise, air pollution, traffic...), social and morphologic differentiations (private buildings, individual houses, social housing...), and rehabilitation projects (areas likely to be re-qualified in the medium term).

The concept of environmental quality of life is then addressed by combining approaches from two different scientific traditions:

- An approach based on measurements and numerical modeling of physical phenomena impacting living quality (microclimate, noise, traffic, air quality), their interactions and impacts.

- A geographical and social approach mobilizing different technics of social surveys to understand how people perceive the environmental quality of their neighborhood, and specially their climatic comfort, in connection with other items of human welfare.

The crucial stage of the research was to conceive a strong interdisciplinary protocol and to carry out in-situ campaigns that combine measurements (urban climatology, acoustics, and air pollution) and surveys (questionnaires) in the three selected neighborhoods. This experimental protocol called 'instrumented and commented walks' was implemented in June 2013 on Marseille, then in October 2013 on Paris, and was replicated three times on Toulouse in winter, spring and summer 2014. During several days and three times per day, some groups, made up of scientists and inhabitants walked along a defined itinerary through the chosen districts. While physical scientists working on urban climate, noise and air pollution were collecting continuous data (leading to temperature, humidity, wind, sound and air pollutants indicators), the social scientists were conducting a survey with the inhabitants through a questionnaire administrated at each relevant place of the itinerary. After each 'Instrumented and commented routes', a focus group was carried out with the same participants to deepen and broaden the issues raised in the questionnaire.

Currently, post-processing and cross-analyzes are undertaken in order to identify and characterize potential relationships between physical measurements and people's feelings and perceptions.

2.2. Simulation Setup

The aim of this work with detailed CFD simulations at the neighborhood level is to support the other studies of air quality, noise and thermal comfort (both objective and subjective). In addition, within the project, mesoscale meteorological simulations have been are also performed by Meteo-France and are used to impose more realistic boundary conditions for the CFD domain. Similar high-resolution CFD simulation are used to study short episodes of local urban climate and air pollution dispersion in the neighborhood of Bordelongue in Toulouse, Porte de Bagnolet in Paris and Saint Marcel in Marseille. Here I present the simulation of the first field experiment which was carried out during June 2013 in the residential areas of Saint Marcel in Marseille.

Mesh and Modeling Strategy

A main challenge, in this case, is the size of the study domain. It has 4 km^2 which is almost 5 times larger than the CAPITOUL one. The domain contains more than 1200 buildings. Moreover, the site is not on flat ground and has significant vegetation areas. It is worth mentioning the geometric modeling and mesh generation in this study.

GIS Dataset Processing. The geographic data is provided by the French Institute of Geography (IGN). The database contains files (shape-file database: digital vector storage format for storing geometric location and associated attribute information) which correspond to different elements of the city. First, the data is processed by using an open-source Geographic Information System (GIS) Quantum GIS (www.qgis.org) to load all raster and vector layers. Since the database was not designed to validate CFD studies, an essential first step was to simplify and optimize the 2D geometry through different sub-steps such as simplifying some forms of buildings, removing smaller geometries according to the criteria that the author set, grouping small houses in the same block, adding missing elements etc (Fig.4.17).

Mesh Generation. After creating a 2D geometry model, the author imports it into the open-source geometry and mesh generator SALOME (www.salome-platform.org). 2D building polygons are extruded vertically to create 3D buildings by using the information provided in building heights. Since the site has a varied topography, the relief is also added, using data from the IGN with a resolution of $25x25m$. The site containing the entire urban structures (buildings, ground type, topography etc.) is then meshed with an automated procedure (Fig. 4.18). The unstructured grid contains about 6 million hybrid meshes ($L \times W \times H : 2 \times 2 \times 1$ km in dimension). The minimum resolution of the grid is $1m$.

Initial and Boundary Conditions

The boundary conditions are obtained either from upper air and ground based measurement from the site during the experiment or from a mesoscale model run by Meteo-France over the entire city at a grid scale of $250m$, which includes an urban parameterization. The timestep has been set to $10s$ for dynamic and $5min$ for thermal-radiative.

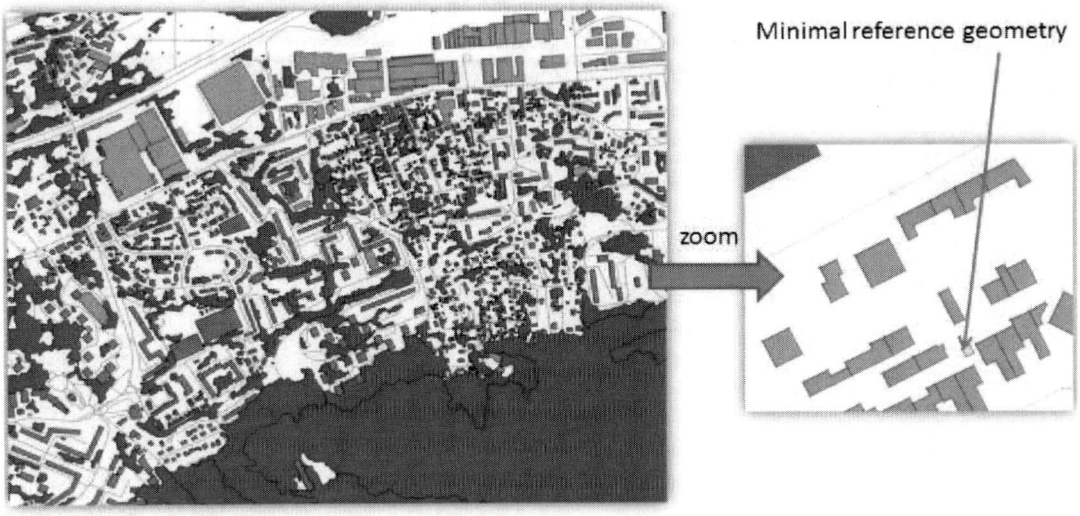

Figure 4.17. 2D urban elements in the database of Marseille, Saint Marcel neighborhood.

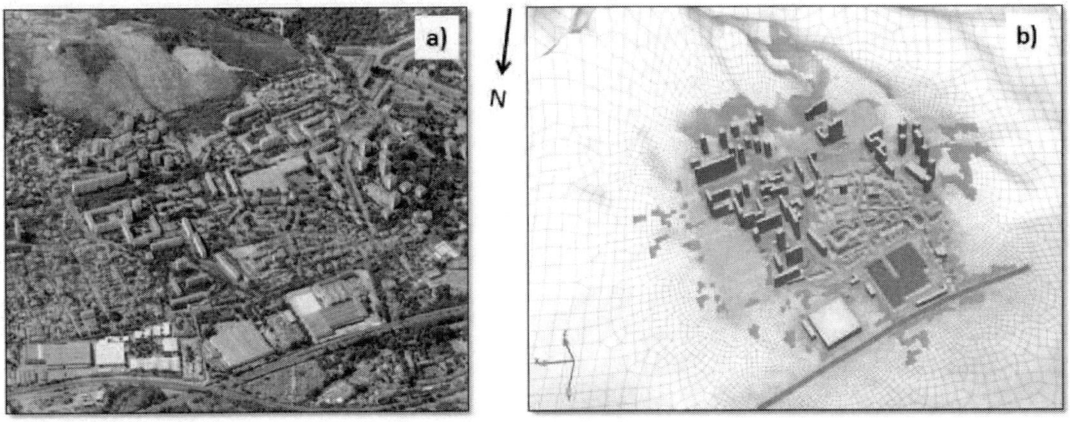

Figure 4.18. Aerial view of Marseille, Saint Marcel neighborhood, France: a) main study area (Bing Map); b) mesh contains buildings, topography, vegetations, highway, ground (identified by their location and properties). The north is indicated by the black arrow. The domain is approximately 2×2 *km*.

2.3. Results and Discussion

The entire 24-hour simulations took a total of 5 hours with 512 processors.

Heat Flux Measurement from Thermal Infrared Imagery

In flat and uniform terrain, the IR fluxes are horizontally homogeneous and isotropic, depending only on the vertical direction. The diurnal evolution, with ground cooling at night and heating during the day is very uniform. This can be modeled with the plane-parallel

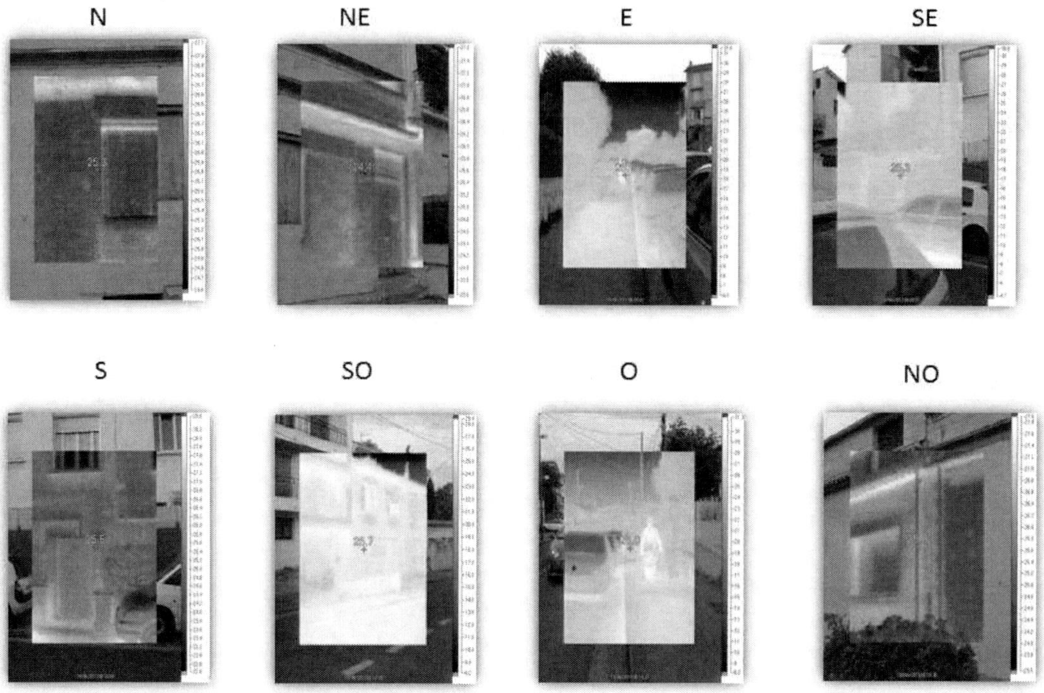

Figure 4.19. Infrared images overlaid on the visible image at the stop point C2 (Fig.4.20) of June 13th, 2013, at 0600 UT following the eight cardinal directions.

approximation where the fluxes depend only locally on the vertical coordinate and that is the basis of a number of radiative models.

In urban areas, however, local variations of the urban fabric lead to both a horizontal variation of the upward IR fluxes and to an anisotropy of their horizontal component. For example in a northern hemisphere mid-latitude city, a southward facing wall receiving the sunlight will radiate more than its north facing counterpart and this will evolve throughout the day and will depend on the local building layout. At night, this will tend to equilibrate with the neighboring buildings. To take these effects into account the radiative model needs to be three-dimensional.

In order to document and quantify these effects experimentally, an IR imagery has been deployed accompanying a mobile meteorological measurement system and sound recordings during EUREQUA project. As mentioned above, these mobile systems were walked in the neighborhood of Marseille approximately every three hours during three days. At each of the 15 stop points for each hourly walk, one ground picture, and eight horizontal pictures have been taken, one for each direction (Fig.4.19). Figure 4.20 displays the walking trajectory (Fig. 4.20a) and the stop points (Fig.4.20b) through numerical modeling.

The imagery records simultaneously a visible picture, an IR picture, and a text file containing the IR measurements that can later be reprocessed. The images then need to be manually tagged to each stop point and direction. These two parts have been the most labor intensive of the data collection. For each hourly walk, more than 100 images have been collected. This means 2500 $\sim$ 3000 images collected for one Intensive Observing Period

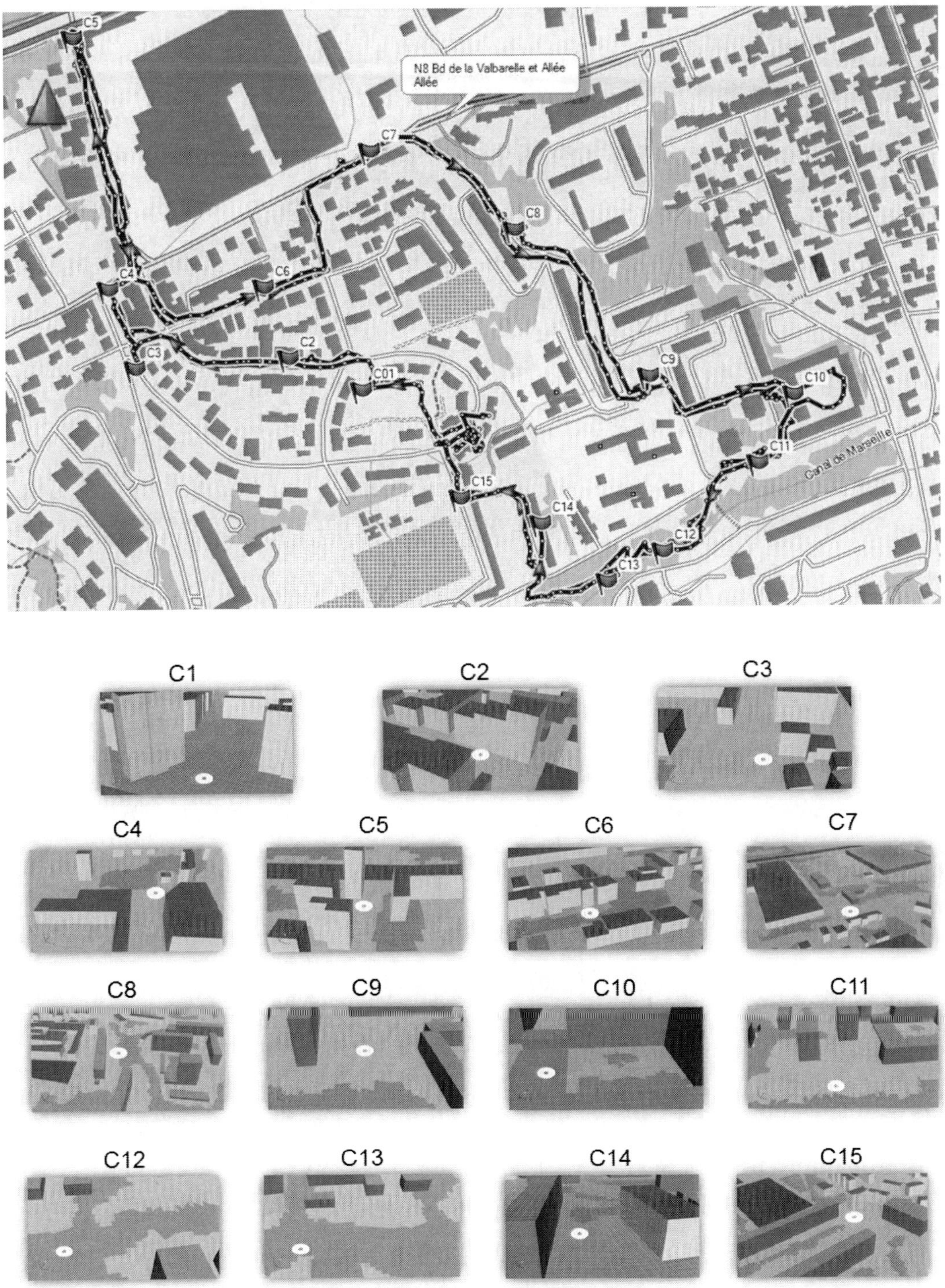

Figure 4.20. Predefined circuit with 15 stop points for mobile sensors (top) and visualization of numerical stop points (bottom).

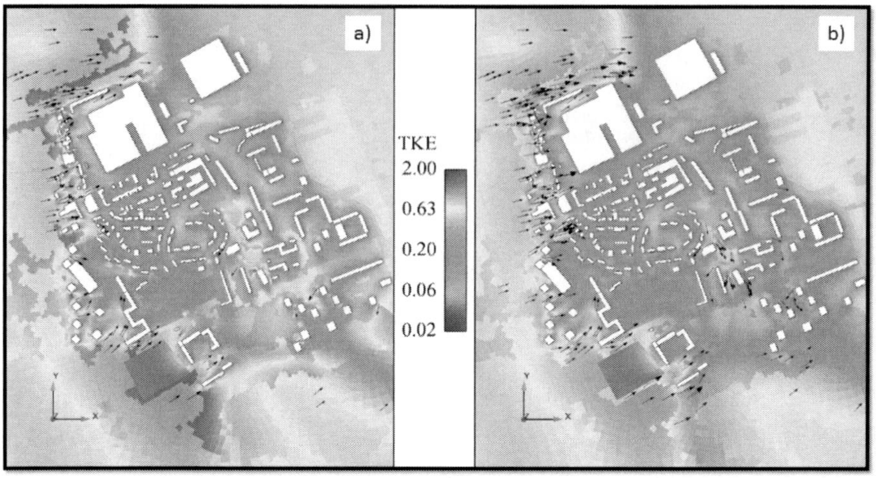

Figure 4.21. Comparison of the mean turbulent kinetic energy (TKE) ($m^2\ s^{-2}$) distribution on ground level: a) Simulation with roughness model; b) Simulation with drag-porosity model.

(IOP) of 3 days, which has been repeated in each field experiment.

Drag-Porosity Model

The canopy creates a drag force against the air flow, it causes a decrease in wind speed and changes the turbulence. In order to get a better estimation of the canopy effect on the wind and turbulence, the non-resolved features such as vegetated area and surrounding non-resolved city area are parameterized by a drag-porosity model (Zaïdi et al., 2013). Thus, the pressure and viscous forces due to the vegetation and surrounding area are modeled by a single force F_i as a source term in the Navier-Stokes equation:

$$F_i = -\frac{1}{2}\rho\alpha C_D|U|\,\overline{U_i} \qquad . \qquad (4.7)$$

where $|U|$ is the local mean velocity magnitude, U_i is the $i\,th$ component of the wind vector and ρ is the air density. The value of the canopy drag coefficient C_D ranges generally between 0.1 and 0.3, depending on the type of trees. In this study, the value of C_D is set to 0.2. α is the leaf area density (leaf foliage area per unit of volume) or building density which has been set to 0.5. The effect on the TKE is modeled with additional terms S_k and S_ε in both the k and ε equations, which is expressed as:

$$S_k = \frac{1}{2}\rho\alpha C_D B_p|U|^3 - \frac{1}{2}\rho\alpha C_D B_d k|U|\,|U| \qquad (4.8)$$

$$S_\varepsilon = \frac{\varepsilon}{k}C_{4\varepsilon}\rho\alpha C_D|U|^3 - C_{5\varepsilon}B_d k|U|\,|U| \qquad (4.9)$$

where $C_{4\varepsilon}$, $C_{5\varepsilon}$, B_p and B_d the model constants. $C_{4\varepsilon}$, $C_{5\varepsilon}$ are both set to 0.9. B_p and B_d are set to 1 and 5.03 respectively.

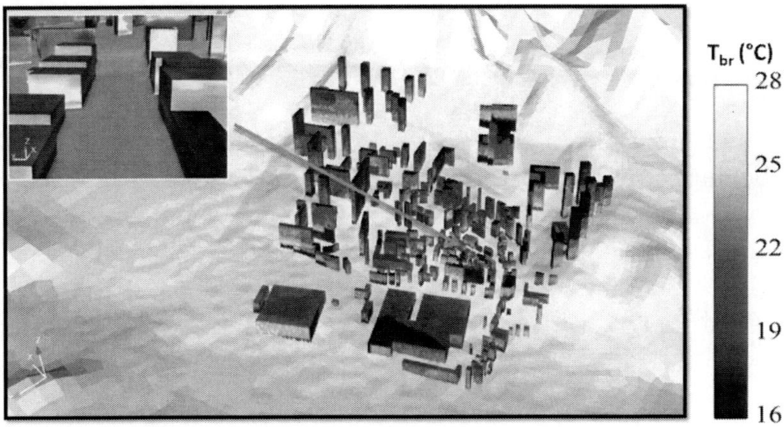

Figure 4.22. Simulated surface temperature brightness for the ground and buildings at 0300 UTC after some nightime cooling.

In order to illustrate the improvement by using the drag-porosity model, the simulated turbulent kinetic energy results are compared between a case using the roughness model (Fig.4.21a) and a case using the drag-porosity model (Fig.4.21b). The relief is not taken into account in this comparison. The wind is blowing from west to east. A high roughness is used for high-density areas. The presence of obstacles slows down the airflow and increases heterogeneity in the TKE. As expected, the drag-porosity model shows a more realistic estimate on the flow dynamics because it slows down the airflow, with a decrease of TKE from the upstream in the vegetation areas where the roughness model shows a increase of TKE.

Thermal Analysis

A diurnal cycle of the atmospheric flow, turbulence and temperature evolution with pollutant dispersion released from the motorway to the north of the domain has been simulated (cf. Chapter 5). In addition infrared temperature images are taken from various points of view, at different locations and in different directions, they are then compared with the corresponding calculation of infrared fluxes. Figure 4.22 displays the simulated brightness temperature for the building walls at 0300 UTC. This temperature represents the blackbody temperature that would give the same flux as the sum of the real emitted and reflected infrared flux and is directly comparable to the infrared camera images that can be obtained from a certain area (e.g., indicated on the top-left zoom in Fig.4.22). This temperature illustrates the surface cooling of the building at night. The surface temperature for building depends on the interior temperature while the ground temperature depends on the deep soil temperature. In these simulations, the deep soil temperature was probably too high and could be corrected by using the average from the previous few weeks.

2.4. Summary

The objective of this work is to perform CFD simulations in the neighborhood of Marseille. A new methodology has been established to generate a mesh for a complex site. The results from the first simulations give a good order of magnitude. Exploiting the field campaign database (containing for example over 2500 infrared and visible images at 15 locations, 8 directions every $3h$) is still an ongoing work in addition to more classical meteorological measurements to validate in more details the thermal-radiative-dynamic modeling.

Chapter 5

Application III: Pollutant Dispersion

1. Introduction

The large population densities that characterize urban areas unavoidably produce a considerable amount of anthropogenic emissions. Common forms of pollutant include the following: hazardous material release, industrial, automobile, and power plant emissions. The challenge of modeling pollutant dispersion is thus understanding how materials are mixed by turbulence and transported from release their point to larger scales (Fernando et al., 2010). Even under ideal conditions in a laboratory the dynamics of turbulence and turbulent diffusion are some of the most difficult in fluid mechanics to model. There is no complete theory that describes the relationship between ambient concentrations of air pollutants and the causative meteorological factors and processes (NIWA, 2004).

According to NIWA (2004), an atmospheric dispersion model is a:

- mathematical simulation of the physics and chemistry governing the transport, dispersion and transformation of pollutants in the atmosphere
- means of estimating downwind air pollution concentrations given information about the pollutant emissions and nature of the atmosphere.

Today most modern air pollution models are computer programs that calculate the pollutant concentration downwind of a source using information on the contaminant emission rate, characteristics of the emission source, local topography, meteorology of the area, ambient or background concentrations of pollutant.

1.1. Gaussian-Plume Models

The most commonly used dispersion models are steady-state Gaussian-plume models. Named for the Gaussian distribution, these are based on a mathematical approximation of plume behavior and are the easiest models to use. Begin with the premise that a turbulent diffusivity, D_t, can be defined and used in an advection-diffusion equation describing plume evolution of the following form:

$$\frac{\partial C}{\partial t} + u.\nabla c = \frac{\partial}{\partial x}(D_{t,x}\frac{\partial c}{\partial x}) + \frac{\partial}{\partial y}(D_{t,y}\frac{\partial c}{\partial y}) + \frac{\partial}{\partial z}(D_{t,z}\frac{\partial c}{\partial z}) \quad (5.1)$$

Information may then be obtained by solving a simplified version of Eq.5.1. In practice these models work from the following basic solution:

$$c(x,y,z) = \frac{Q}{\pi U \sigma_y \sigma_z} exp[-\frac{1}{2}(\frac{y^2}{\sigma_y^2} + \frac{z^2}{\sigma_z^2})] \tag{5.2}$$

where Q is the release rate of pollutant per unit time and U is a uniform advection velocity (in y and z). σ_y and σ_z are the horizontal plume width and vertical plume height, respectively. Formally they are related to the turbulent diffusivity from Eq.5.1 by the expression $\sigma_i = 2xD_{t,i}/U$, where x is the distance from the source. Since the turbulent diffusivity is unknown, however, the plume width and height are parameterized based on experimental measurements and observations. The parametrization is typically a function of conditions like atmospheric stability (Holmes and Morawska, 2006). U may also be specified as an increasing function of x to approximate the effects of vertical wind shear (Pasquill and Smith, 1983). As far as urban areas are concerned, the roughness length concept is simply used to adjust the parametrization of U or the plume spreading rates. However, even advanced Gaussian models will only accept a single value for surface roughness.

1.2. CFD Models

If a Gaussian-plume model is inappropriate for any particular application because of its limitations, and a full puff model meteorological data set is not available, an advanced model with a single-point meteorological data set should be considered.

The advanced dispersion models used avoid most of the limitations associated with steady-state models, such as CFD models. As discussed previously in Chapter 2, a CFD model refers to solving the Navier-Stokes and scalar transport equations numerically in a geometry divided into discrete computational cells. The equations enforce basic mass and momentum conservation for the fluid (air) and scalars (pollutants) involved. Modeling enters into the process when turbulent motions occur on length and time-scales that become too computationally demanding to resolve. The three fundamental CFD approaches discussed in Chapter 2, DNS, LES, and RANS vary in the amount and type of turbulence parametrization. A key advantage of all CFD models as compared to the Gaussian-plume models is the ability to more effectively capture the effects of complex geometry on the dispersion process. Geometry enters a CFD model naturally as a boundary condition, rather than through empirical parametrization of an analytical solution as in a Gaussian-plume model. Although their demands on resources (human, computational and data) are far higher than those of Gaussian-plume models, computer power is also increasing rapidly, making this aspect less of an issue. However, the use of CFD models does involve much greater meteorological input data demands.

Still, the CFD approach is not without its own modeling assumptions. The most assumptions in all computations that merit discussion are that of passive scalar transport and neutral stability. A passive scalar transport means that a given pollutant does not react in the underlying flow that is transporting it. Atmospheric stability is a measure of the propensity for vertical motion and hence is an important indicator of the likely magnitude of pollutant dispersion. In neutral conditions, a plume will tend to be uplifted by broad terrain features. As it relates to atmospheric dispersion, neutral stability is a more limiting assumption. For the dispersion scenarios and analysis presented in this Chapter, the passive dispersion is an acceptable approximation.

2. Numerical Study of Thermal Impact on the Dispersion of a Passive Pollutant in Complex Urban Areas

In contrast to the previous chapter where the developed thermal-radiative-dynamic coupling as a tool for urban Surface Energy Balance simulation was validated, in this chapter CFD computations are performed to study dispersion physics with the real urban building canopy.

2.1. Example for the City Center of Toulouse

CFD simulations in the Toulouse city domain were performed in this study to investigate the thermal effects of buildings on dispersion (Qu et al., 2013). Hazardous waste was released at ground level, starting at 1000 UT for a period of 3 hours as shown in Fig.5.1 (white point).

Simulation Setup

Open source CFD code *Code_Saturne* is employed to carry out the simulation. The dry atmosphere option is activated solving the Navier-Stokes equations for the air flow and the transport equations for the potential temperature so as to take into account the thermal effects. The modified $k-\varepsilon$ model is used for the turbulence modeling (see Chapter 2). The simulation domain and mesh are the same as in the CAPITOUL case (Fig. 4.3). The meteorological initial and inlet boundary conditions are taken from a neutrally stratified wind profile blowing westward with a reference 47.5-m wind speed $U_{ref} = 3\ m\ s^{-1}$, associated with a neutral potential temperature profile of 22°C (Fig. 5.1).

In order to study the contribution of atmospheric radiation on the dispersion, two cases have been performed: a reference case without heat transfer, i.e., with a neutral stratification of the atmosphere, called hereafter 'neutral case', and a case with the 3D atmospheric thermal-radiative model coupled to the dynamical one called hereafter 'thermal-radiative transfer case'.

Results and Discussion

The numerical simulation results are analyzed at the end of the release when the mean flow in the entire simulated urban canopy reaches a quasi-steady state in the neutral case. Airflow in the urban canopy is composed of very complicated vortices rotating in the horizontal and vertical directions. Figures 5.2a and b show the TKE fields on the vertical and horizontal cross sections of respectively the neutral and thermal-radiative cases. Compared to the neutral case, production of turbulence by the building is complex (Fig. 5.2a), the presence of heat transfer (Fig. 5.2b) enhances the TKE in the whole domain due to thermal production, in particular near the roof level.

Figure 5.3 shows hot spots of dimensionless pollution concentration near the ground for two thermal conditions. From the neutral case (Fig.5.3a), emission from the ground disperses by the dominating wind along the west-east direction. Since the wind velocities are reduced dramatically inside the canyons, stagnation areas are formed there. This kind of uneven distribution of wind velocities will definitely influence the pollutant dispersion. Hence, in the yard pollutants were both likely less diluted and more concentrated than on the roof (higher yard pollutant concentrations). With thermal transfer condition (Fig. 5.3b), a lower concentration is found downstream, compared to the neutral case, surface heating induces greater dispersion. Inside-yard concentration is lower.

To clarify the change in the dispersion characteristics with changes in thermal condition, the concentration distribution is plotted on a vertical cross section at the released source position in Fig.5.4. With thermal transfer conditions (Fig.5.4b), the air temperature is increased which results in an upward motion. Buoyant forces are dominant in this low-wind scenario, as the plume is seen to trend upwards from the point of emission. Moreover, the chosen thermal conditions do lift the plume sufficiently away from the surrounding buildings. Therefore, higher concentrations are found above the road in comparison with the neutral case (Fig. 5.4a).

2.2. Example for the Neighborhood of Marseille

Having established the process by which a dispersion simulation in an urban environment can be constructed and run from the previous section, this section is concerned with the numerical results to EUREQUA project for the area Saint Marcel in Marseille (cf. Chapter 4 section EUREQUA). More specifically, this section will predict air pollutant concentrations near road-ways CFD simulation for the area Saint Marcel in Marseille.

Simulation Setup

Details of the numerical methods of the simulation can be found in Chapter 2. Besides the general setup, a drag-porosity model is introduced in the code to represent the effects of the vegetation area and surrounding urban canopy layer (cf. Chapter 4 section EUREQUA).

The meteorological profiles obtained from the code Meso-NH performed by Meteo-France over the region were used as boundary conditions at the edges of the simulation domain. Due to the coarse grid used in mesoscale simulation, the meteorological data were interpolated in time and in space to fit the temporal and spatial resolutions in this case. The spatially interpolated meteorological data at the first instance is used as the initial condition for the flow inside the domain. A zero Neumann condition is applied at the outlet boundary. On the solid surfaces, a rough wall law is defined by setting a zero flux condition for the scalars.

The pollutant is considered as passive scalar, transported by local winds and dispersed by local turbulence, which are solved by the transport equation. The local air pollution emissions are principally due to traffic on the highway and local streets, which pass through the neighborhood. A source term added to the first grid's layer where the cells are next to the surfaces boundary of the highway (Fig. 4.18). Traffic emission is supposed to be a linear function of traffic volume: every 100 vehicles correspond to 1 $mg\ m^{-3}$ emission.

Results and Discussion

Figure 5.5a illustrates the dimensionless concentration at 2 m at 1100 UT for a neutrally stratified wind profile blowing southeastward. The high traffic volume increases vehicles emissions during the day. Significant concentration can be found near the highway. The wind direction (black vectors in Fig.5.5) greatly influences the airflow and dispersion patterns above the canopy, through the forests and across the streets. Due to the thermal transfer condition, higher air temperature can be observed close to the buildings in Figure 5.5b. As explained in the previous section, surface heating influences the local flow field and therefore induces greater local dispersion. Overall, traffic-related emissions are a key contributor to the concentration in areas with high road traffic volumes.

2.3. Summary

Using new atmospheric radiative and thermal schemes implemented in *Code_Saturne*, the thermal influence has been tested on pollutant dispersion. The simulation results show not only the interest of using a CFD code to predict the dispersion within complex geometries but also that the thermal effects considerably alter the plume shape.

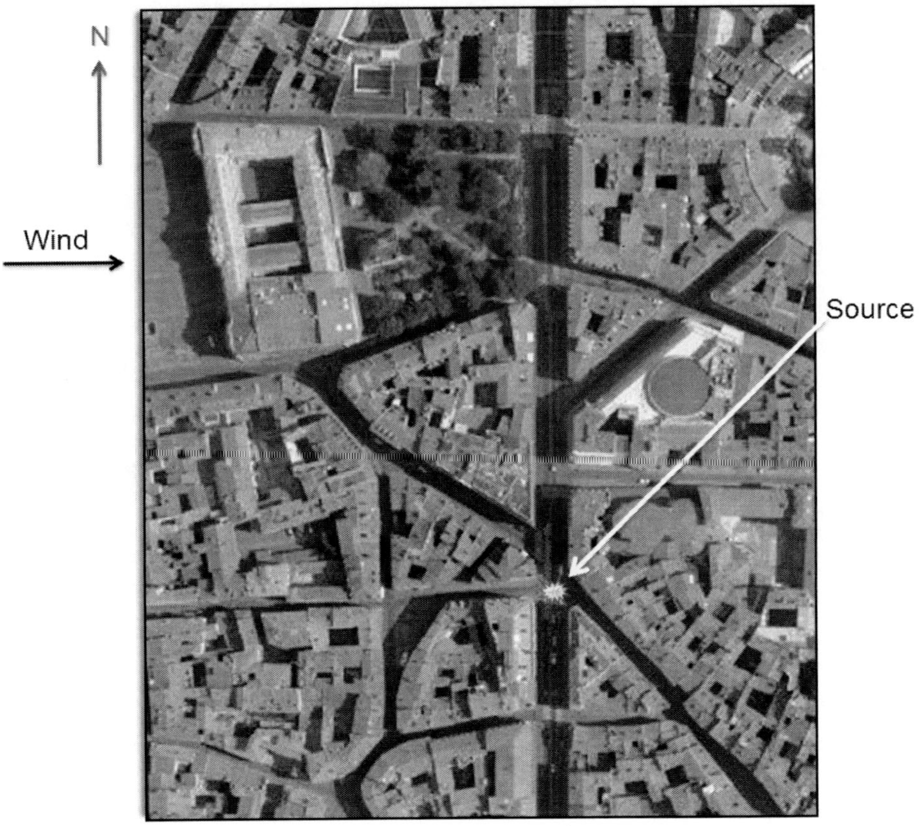

Figure 5.1. Aerial view of main study area, from Google Maps. The wind direction and north are indicated.

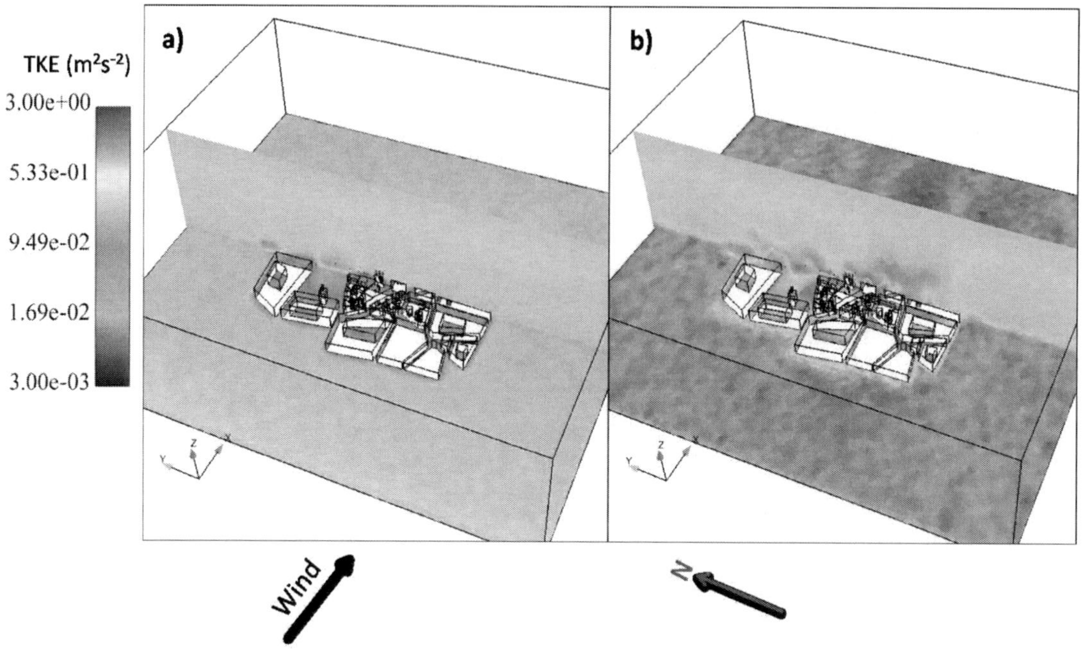

Figure 5.2. Comparison of the mean Turbulent Kinetic Energy (TKE) ($m^2\ s^{-2}$) distribution on the vertical- and horizontal cross sections: a) neutral; b) thermal-radiative transfers. The wind direction and north are indicated.

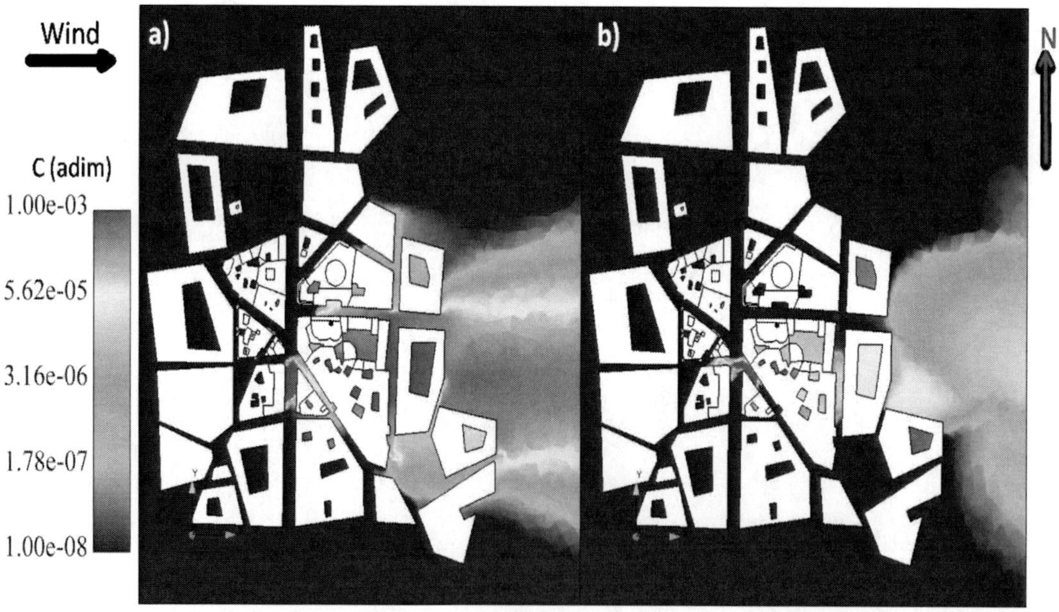

Figure 5.3. Comparison of the concentration distribution on the ground level at the end of the release: a) neutral conditions; b) thermal-radiative transfer conditions. The wind direction and north are indicated.

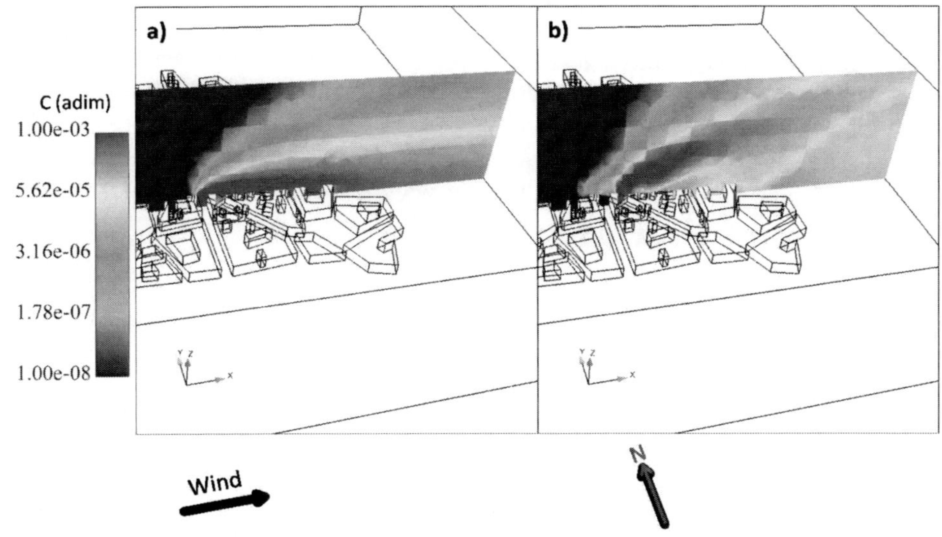

Figure 5.4. Comparison of the concentration distribution at the end of the release on a vertical cross section where the emission released: a) neutral conditions; b) thermal-radiative transfer conditions. The wind direction and north are indicated.

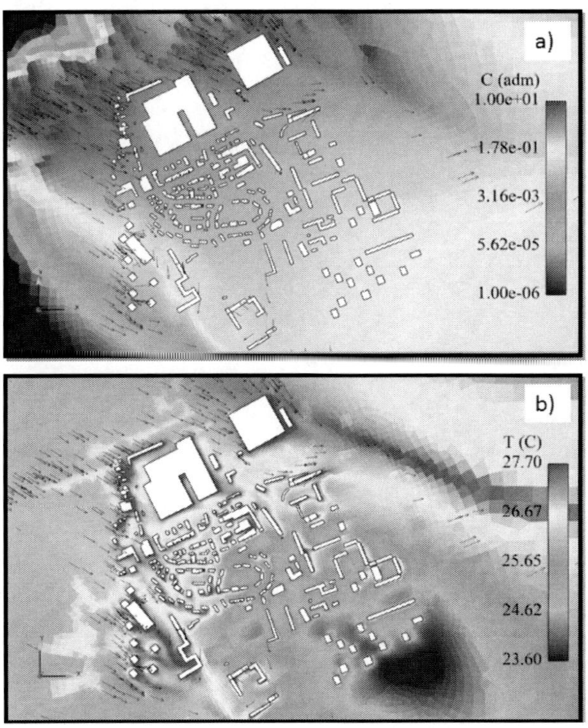

Figure 5.5. Dispersion simulations on Saint Marcel of Marseille city mesh at 1100 UT; visualization of the horizontal cross section on: a) dispersion patterns; b) air temperature.

Chapter 6

Conclusion

"A model is a simplified picture of reality. It does not contain all the features of the real system but contains the features of interest for the management issue or scientific problem we wish to solve by its use."

– National Institute of Water and Atmospheric Research, New Zealand

1. Results and Conclusion

In this work, a three-dimensional numerical atmospheric radiative transfer scheme coupled with a computational fluid dynamics code to simulate the urban Surface Energy Balance (SEB) and pollutant dispersion in complex geometries has been developed (Fig.6.1). Validations of the full thermal-radiative-convective coupling were performed through a comparison with observation datasets. The assessment of numerical modeling approaches began in fundamental idealized built-up area with simplified building block geometries (MUST field experiments, Chapter 3) then in real urban environments with complex geometries (CAPITOUL and EUREQUA projects, Chapter 4). Additionally, full-scale simulations in Toulouse and Marseille cities were performed to aid in integrating tools for 3D model construction, automated meshing, and CFD simulation.

The tools developed in this work can be applied to detailed studies of local urban climate, to the study of present conditions and future scenario, such as the densification of housing and the introduction of green area or green/white roofs. It can be used to further assess the street canyon ventilation potential, the possible shading strategies on building surfaces and the influence of both aspects on indoor thermal comfort and air quality. This tool is also well suited to the study of 'hot spots' in the air quality of urban centers where the emissions are particularly concentrated and to evaluate the alternatives to mitigate them. It can also contribute to future research and applications in the field of wind engineering in the urban environment, when the thermal stratification is of importance.

This chapter will recap the major findings relating to the numerical methods investigated and comment more broadly on how they fit into modeling SEB and a hazardous urban release scenario.

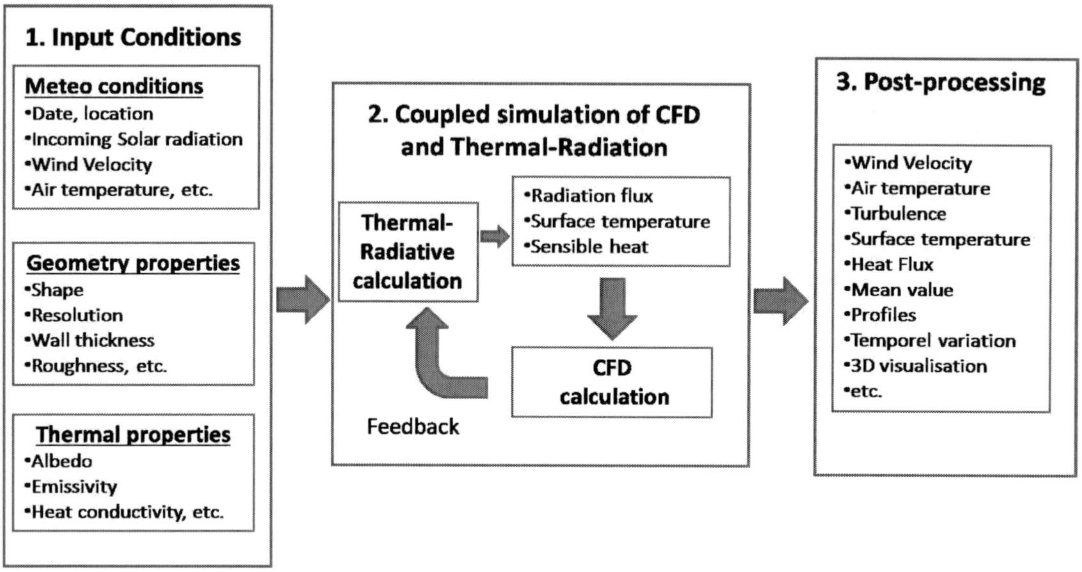

Figure 6.1. Algorithm outline of the modeling approach.

1.1. Thermal Transfer Model

The key to the evaluation of urban SEB models was the accuracy of surface temperatures. The improvement of the thermal transfer model for buildings and the study of the full thermal-radiative-convective coupling, were addressed in Chapters 3 and 4. The model successfully simulated the surface temperature, for different sides of a container within the MUST canopy in a diurnal cycle (cf. Chapter 3) and also a 24-hour evolution of the brightness surface temperature for different positions in the district within the CAPITOUL study center area (cf. Chapter 4). The model incorporates two approaches to model heat transfer with buildings: the force-restore scheme and the wall thermal model. Tests of the performance of these two approaches showed that to explicitly treat the heat conduction, even with only one-layer, the wall thermal model was more adapted for the building walls than the force-restore approach which was originally developed for the soil model (cf. Chapter 3). By taking advantage of the force-restore approach and the wall thermal model to separately model the ground and building walls temperatures, a hybrid surface temperature scheme showed a good ability when applied to the CAPITOUL and EUREQUA cases (cf. Chapter 4). In case of buildings without good insulation, the internal building temperature had a large influence on the surface temperature in the diurnal cycle (cf. Chapter 3). As a solution, an evolution equation obtained from the literature was adopted to more realistically model the internal building temperature.

1.2. Radiation Model

Based on MUST full coupling validation, the atmospheric radiation model in *Code_Saturne* was well-tested in Chapter 3 to achieve the objective of comparison with the geometric view factor approach, used by the SOLENE model. The most significant differences between the two models were for the diffuse solar flux, due partly to an assumption of the homogeneous incoming diffuse solar flux from simple clear sky model in *Code_Saturne*. Nevertheless, results showed that the two models agreed well with each other for the time evolution of the direct solar flux, incident infrared flux and surface temperature, because the diffuse solar fluxes were relatively small in clear sky condition.

1.3. Dynamic Model

The presence of natural or artificial asperities (buildings, hills, trees, vegetation, water) creates an aerodynamic drag against the flow, which causes a decrease in wind speed and modifies turbulence. However, it is difficult to take into account explicitly all these features in the mesh for the high-resolution CFD study. In the CAPITOUL case (cf. Chapter 4), the buildings in the region outside were treated with a high roughness value. In order to improve the dynamic model, a drag-porosity approach was applied in the EUREQUA case (cf. Chapter 4) to retain a more realistic parametrization for the surrounding urban canopy layer and for the vegetation zones. A comparison between the drag-porosity model and the classical roughness law confirmed that the drag-porosity model gave a more realistic estimate on the turbulent kinetic energy.

1.4. Computational Infrastructure for Urban CFD Simulation

A running theme of this monograph was integrating a set of tools developed in a research context to apply to urban CFD simulation. Accomplishing this objective requires modeling a real city's urban topography. The automated meshing tools described in Chapter 4 are designed to work with 3D real urban geometries. The generation of the CAPITOUL complex mesh required an important preliminary work to optimize the geometry, and in addition, 4 classes of albedo value have been separated depending on wall painting colors. The example in the EUREQUA case described a second automated meshing tool utilizing GIS data to construct the residential areas of Saint Marcel in Marseille. The process is repeatable for any urban geometry characterized by a CAD or GIS dataset. The flexibility of these meshing tools has proved very useful in preparing simulations of SEB and of dispersion.

1.5. Thermal Analysis

As a demonstration of the ability of the thermal-radiative model, the thermal effect of buildings on the local atmospheric flow under low wind speed condition has been also investigated in Chapter 3. In this short-term study, the results showed that taking into account the thermal stratification had a significant influence on the wind field. The difference between the case for which the temperatures were imposed and the case with temperatures calculated with 3D radiative transfer case could also be important. Moreover, under radiative condition, the temperature and vertical motion showed a great sensitivity to change in the physical parameters of the wall surfaces. Increasing the albedo from 0.1 to 0.6 (e.g., painting a dark surface to white) reduced the net radiation flux term in SEB, as a result, the temperature close to the windward wall or roof might decrease by 2 °C.

The results from the CAPITOUL case described in Chapter 4 have addressed one of the most important objective: the validation of full coupling approach on a real city district. Through the comparison with the Thermal InfraRed airborne images from two flights in the day of 15th July 2004, the importance of taking into account heterogeneities in materials and geometry were demonstrated to reproduce the spatial variability of the temperatures in complex urban areas. Beyond being just a validation set for surface temperatures' computations, the heat fluxes were examined to explore the SEB in CAPITOUL case. The full coupling was tested against in situ observation during the same day. The model predicted a good daytime sensible flux, but showed a nighttime underestimation at roof surface. The reason to this underestimation can be explained by taking into consideration the uncertainty from the measurements and non-detailed geometries of the roofs. Similar comments can be made for the model-observation comparison of the outgoing infrared flux. Regarding the outgoing solar flux, the result showed an excellent model-observation agreement at the mast position.

Results from Chapter 3 and 4 also illustrated quantitatively how the sensible heat flux played an important role in SEB. Moreover, in Chapter 3, different approaches have been discussed and been compared to estimate the sensible flux: constant heat transfer coefficient, using a 1D equation based on a vertical wind profile within the canopy and the 3D modeling approach. This comparison shows a difference about 4 °C on the wall. Thus it can be concluded that a good description of the heat transfer coefficient and its spatial variability is essential to determine the sensible heat flux in the canopy.

1.6. Pollutant Dispersion

Moving to scalar dispersion applications (Chapter 5), the evaluation of CFD modeling of dispersion focused generally on the anisotropies present in complex geometries. The grid resolution requirements was leveraged to simulate several dispersion scenarios, care was taken to ensure accuracy with previously established CAPITOUL and EUREQUA SEB simulations. The flow configuration was selected to further study the effects of a hazardous waste or traffic emission within the canopy. The instantaneous plume structure was examined to different atmospheric assumptions. In neutral atmospheric conditions the pollutant tended to be well mixed in the street canyons. However in stratified conditions, the thermal effects considerably altered the plume shape.

2. Remaining Challenges

There are still a number of directions avenues which remain to be investigated or which have opened during this work. The numerical simulations were carried out with the CFD code *Code_Saturne*. The approaches developed in this work should not be limited only to testing *Code_Saturne*. They should be easily implemented into other open-source CFD codes.

For the full-scale CFD simulation of Saint Marcel in Marseille city, mesh generation process can benefit from further automation in more complex geometries (e.g., Paris and Toulouse cities in the EUREQUA project).

Taking Toulouse city once again as an example, the application of the thermal-radiative-dynamic coupled model could be extended through more model-observation comparisons, in particular for a winter day. If the anthropogenic heat flux is not significant in a summer day, it may not be neglected in winter. Thus, the model should properly include variation of diurnal anthropogenic flux, eventually by using a simple formula proposed by Pigeon et al. (2008). Moreover, the comparison of model results with measurements carried out in winter day would address in a different way the issue of the condition of internal building temperature.

The one-layer wall thermal model showed some limitations. In MUST case, it reproduced the diurnal cycle of the different surface temperature only with a very low thermal conductivity. Either in MUST or in CAPITOUL case, a nighttime overcooling appears by using this scheme. At day scale, at least two layers for a wall scheme are requested to obtain the surface temperature rapid changes and diurnal cycles (i.e., heat storage) (Dupont and Mestayer, 2006). The implementation of a multilayer wall thermal model may lead to improvements and can improve the temperature underestimation during the night.

The comparison of *Code_Saturne* and SOLENE radiation models and the numerical study of a low wind speed case in Chapter 3 may be supplemented with more real measurement data set. Furthermore, when comparing the two radiation models, taking diffuse solar flux values from SOLENE as boundary conditions showed an improvement in *Code_Saturne*. Full implementation of a similar sky model as in SOLENE with non-uniform diffuse solar radiation distribution may be necessary to *Code_Saturne* when applied under cloudy weather conditions. The comparison also can be extended to other field experiments, such as with results presented by Hénon et al. (2012) from CAPITOUL datasets, or to compare with other radiation models, e.g., SYRTHES (SYstèm de Résolution THErmique Solide) (Péniguel and Rupp, 2008). SYRTHES is a thermal code for a larger number industrial of applications and has also become open-source. The thermal solver can be used alone if only by conduction/radiation problems or coupled with *Code_Saturne* or other fluid mechanics codes.

Thermal effects of the buildings and ground have not been included to study the dispersion in Toulouse and Marseille cities but the numerical results were not compared to the measurement yet. Validation with experimental data will remain indispensable, and should continue to accompany further developments in and applications of the model.

In addition, the drag-porosity model was used in the EUREQUA case to improve the dynamic model for the vegetation zones. Other additional improvements for the results might be considered such as estimating daily net radiation over vegetation canopy especially in the EUREQUA case. The amount of radiation on the soil depends on the structure of the vegetation above, which determines the canopy openness, and the proximity of open spaces in the forest, e.g., forest gaps. The openness of the canopy differs per forest type and depends to a large extent on leaf area. A function given by Monsi and Saeki (1953) for the radiation extinction through the vegetation could be used to improve the interaction of radiation with vegetation and soil.

In the model presented in this work, the atmosphere at the microscale is considered to be transparent and non-diffusive. One of the perspectives of this work is to use the 3D radiative scheme for nontransparent media, for applications like assessment of pollutant dispersion or fog formation and dissipation etc. Absorption and diffusion have significant importance. Absorption can already be taken into account by our scheme, so it requires adding the diffusion term in the resolution of the radiative transfer equation.

Regarding the dynamical aspect, in the present work, the turbulence is parameterized by the Reynolds-Averaged Navier-Stokes (RANS) (more specifically, the standard $k-\varepsilon$

closure) approaches due to the limitations on the computational cost. It is well known that conventional $k-\varepsilon$ models, with the turbulent viscosity formulation, tend to overestimate the level of turbulent kinetic energy around the stagnation point. In order to increase the accuracy of the prediction of the airflow around buildings, one direction of future work is to use the Large Eddy Simulation (LES) approach for the convective model. A comparison between LES and RANS computations for the study of contaminant dispersion in the MUST field experiment has been assessed by Santiago et al. (2010). At the microscale level, small irregularities are shown to affect significantly the mean vertical velocity component while the mean stream-wise velocity and Reynolds shear stress are shown to be less sensitive to small geometrical perturbations. For the mean streamwise velocity and Reynolds shear stress, the present LES results are found to be close to RANS results and both approaches were in satisfactory agreement with the observations. However, LES captured better the irregularity effects observed on the vertical velocity components. The magnitude of this velocity component is in general underestimated by RANS. At the mesoscale level, the small geometrical perturbation effects were found insignificant for both the spatially-averaged stream-wise and vertical velocity components and as well for the spatially-averaged Reynolds shear stress. Regarding the dispersive stress, it was shown to be negligible compared to the spatially-averaged shear stress. Globally, the results obtained with LES and RANS for the spatially-averaged flow properties were found to be similar for each flow configuration considered and only slight differences were observed in their cases studied (LES in regular and irregular arrays, and RANS in regular and irregular arrays). At this scale level, it was shown that the flow properties averaged over the full MUST array flow configuration are similar to the flow properties averaged over the one unit regular container flow configuration. This result is very relevant from the urban canopy modeling point of view because the spatially-averaged flow properties computed by CFD models in a simplified configuration can be representative of the average properties of a real part of a city without large irregularities, and can be used for the improvement of the parameterization of atmospheric mesoscale models.

In the CAPITOUL case, the simulated sensible heat flux at the roof surface does not compare well with the measured flux on the mast. This is probably due to the comparison not taking into account the micro-meteorology. One solution is to use a 3D back-trajectory model to identify all the surfaces (not only the roof) which are influenced by the convective flux at the top of the mast. However, due to the inhomogeneities at micro-scale, back-trajectory analysis is very difficult in complex geometry with the unsteady 3D field. So another solution is to calculate the convective flux from the simulation of the turbulent fluctuations by using LES model. If the computational cost remains too high for the whole day simulation, at least, LES can be applied for the shortened study such as the low wind speed case described in Chapter 3.

Finally, tests of the full model are limited to three days at three separate sites, and should be performed for a greater range of urban sites, weather conditions, and seasons to more successfully separate model weaknesses from incorrect parameter specification and/or processes influencing the observations. For instance, the Joint ATREUS-PICADA Experiment (JAPEX) in situ experimental campaign of which the buildings were idealized by the containers was established. Two principal objectives of the JAPEX experiment were to evaluate building-street-atmosphere thermal transfers and the energy budget of a street, to document the re-circulation flow in a street (Idczak et al., 2007, 2010). Since the measurement includes more data than in the MUST filed experiment such as radiation data, it can be used to be a complementary work for the studies with the idealized cases (cf. Chapter 3). Another field experiment, Towards Optimisation of urban-planning and architectural Parameters for Energy Use minimization in Mediterranean cities (TOPEUM) campaign has been carried out in the city of Nicosia, Cyprus in July 2010 (Neophytou et al., 2011). The main objective of TOPEUM is to investigate the influence of different urbanization characteristics (such as the geometry and density of buildings within a city and the type of building materials used) on the intensity of the urban heating effect in the case of a typical Mediterranean city, such as a Cypriot urban area. The field measurements include meteorological measurements as well as on-ground and aerial thermal infrared pictures, covering a range of spatial scales, from local-street canyon to mesoscale. The measurements record the meteorology, the resulting local micro-climate, particularly in the street, and the thermal response of the buildings in the field area, especially the building-inside temperature measurement. This can provide a good verification for the accuracy of the evolution equation which was used to estimate the internal building temperature.

References

AIJ, 2004: Recommendations for loads on buildings. Technical report, Architectural Institute of Japan.

Archambeau, F., N. Méchitoua, and M. Sakiz, 2003: *Code_Saturne*: a Finite Volume Code for the Computation of Turbulent Incompressible Flows - Industrial Applications. *Int. J. Finite. Volumes*, 1, 1–62 [Available online at http://www.latp.univ–mrs.fr//IJFVDB/saturne.pdf].

Arnfield, A. J., 2003: Two decades of urban climate research: A review of turbulence, exchanges of energy and water, and the urban heat island. *Int. J. Climatol.*, 23 (1), 1–26.

Asawa, T., A. Hoyano, and K. Nakaohkubo, 2008: Thermal design tool for outdoor space based on a numerical simulation system using 3D-CAD. *Build. Environ.*, 43 (12), 2112–2123.

Baldwin, B. S. and H. Lomax, 1978: Thin layer approximation and algebraic model for separated turbulent flows. *AIAA*, 78–257.

Best, M. J., C. S. B. Grimmond, and M. G. Villani, 2006: Evaluation of the urban tile in MOSES using surface energy balance observations. *Bound.-Layer Meteor.*, 118 (3), 503–525.

Biltoft, C. A., 2001: *Customer report for mock urban setting test*. DPG Document WDTC-FR-01-121, West Desert Test Center, U.S. Army Dugway Proving Ground, 58 pp., Dugaway, Utah.

Bird, R. E. and R. L. Hulstrom, 1981: A simplified clear sky model for direct and diffuse insolation on horizontal surfaces. *Tech. Rep.* SERI/TR-642-761, Solar Research Institute, 38 pp., Golden, Colorado.

Blocken, B., 2009: Pedestrian wind comfort around a large football stadium in an urban environment: CFD simulation, validation and application of the new Dutch wind nuisance standard. *J. Wind Eng. Ind. Aerod.*, 97 (5-6), 255–270.

Blocken, B., T. Defraeye, D. Derome, and J. Carmeliet, 2009: High-resolution CFD simulations for convective heat transfer coefficients at the facade of a low-rise building. *Build. Environ.*, 44 (12), 2396–2412.

Blocken, B., T. Stathopoulos, and J. Carmeliet, 2007: CFD simulation of the atmospheric boundary layer: wall function problems. *Atmos. Environ.*, 41(2), 238–252.

Boussinesq, J., 1877: Essai sur la theories des eaux courantes. *Memoires presentes par divers savants a l'Academic des Sciences de l'Institut National de France*, 23, 46–50.

Bouyer, J., 2009: Modélisation et simulation des microclimats urbains: Étude de l'impact de l'aménagement urbain sur les consommations énergétiques des bâtiments. *Ph.D. thesis*, École Polytechnique de l'Université de Nantes, 306 pp., [in French].

Briggen, P. M., B. Blocken, and H. L. Schellen, 2009: Wind-driven rain on the facade of a monumental tower: numerical simulation, full-scale validation and sensitivity analysis. *Build. Environ.*, 44 (8), 1675–1690.

Brook, D. R., N. V. Beck, C. M. Clem, D. C. Strickland, I. H. Griffits, D. J. Hall, R. D. Kingdon, and J. M. Hargrave, 2002: Validation of the Urban Dispersion Model(UDM). *Eighth Int. Conf. on Harmonisation within Atmospheric Dispersion Modelling for Regulatory Purposes*, Sofia, Bulgaria (October 14-17 2002).

Brown, M. J. and M. D. Williams, 1998: An urban canopy parameterization for mesoscale meteorological models. *AMS 2nd Urban Environment Conf.*, Albuquerque, New Mexico (November 1998).

Camelli, F. E., R. Lohner, and S. R. Hanna, 2005: VLES study of MUST experiment. *43rd AIAA Aerospace Sciences Meeting and Exhibit*, Reno, Nevada.

Camuffo, D. and A. Bernardi, 1982: An observational study of heat fluxes and the relationship with net radiation. *Bound.-Layer Meteor.*, 23 (3), 359–368.

Casey, M. and T. Wintergerste, 2000: Best Practice Guidelines, ERCOFTAC Special Interest Group on Quality and Trust in Industrial CFD. *Tech. rep.*, Fluid Dynamics Laboratory Sulzer Innotec, ERCOFTAC, Brussels.

Chapman, D. R., 1979: Computational aerodynamics development and outlook. *AIAA*, 17, 1293–1313.

Chen, H., R. Ooka, H. Huang, and T. Tsuchiya, 2009: Study on mitigation measures for outdoor thermal environment on present urban blocks in Tokyo using coupled simulation. *Build. Environ.*, 44 (11), 2290–2299.

Cionco, R. M., 1965: A mathematical model for air flow in a vegetative canopy. *J. Appl. Meteor.*, 4, 517–522.

Coceal, O., A. Dobre, T. G. Thomas, and S. E. Belcher, 2007: Structure of turbulent flow over regular arrays of cubical roughness. *J. Fluid Mech.*, 589, 375–409.

Cook, M., Y. Ji, and G. Hunt, 2003: CFD modelling of natural ventilation: combined wind and buoyancy forces. *Int. J. Ventilation*, 1, 169–180.

Courant, R., K. Friedrichs, and H. Lewy, 1967: On the partial difference equations of mathematical physics. *IBM Journal*, 215–234.

Dandou, A., M. Tombrou, E. Akylas, N. Soulakellis, and E. Bossioli, 2005: Development and evaluation of an urban parameterization scheme in the Penn StateNCAR Mesoscale model (MM5). *J. Geophys. Res.*, 110, D10 102.

Davidson, P. A., 2004: *Turbulence - An Introduction for Scientists and Engineers*. Oxford University Press, 678 pp.

Deardorf, J. W., 1978: Efficient prediction of ground surface temperature and moisture with inclusion of a layer of vegetation. *J. Geophys. Res.*, 83, 1889–1903.

Douce, A. and N. Méchitoua, 2003: Mise en œuvre dans *Code_Saturne* des modélisations physiques particulière, Tome 3: Transfert thermique radiatif en milieux gris semi-transparents. *Rapport interne EDF*, [in French], EDF R&D, Fluid Mechanics and Heat Transfer Department.

Dupont, S., I. Calmet, and P. G. Mestayer, 2002: Urban canopy modeling influence on urban boundary layer simulation. *Fourth Symp. on Urban Environment*, Norfolk, Virginia (May 20-24, 2002), pp. 151-152.

Dupont, S. and P. G. Mestayer, 2006: Parameterization of the urban energy budget with the submesoscale soil model. *J. Appl. Meteor. Climatol.*, 45 (12), 1744–1765.

Dupont, S., T. L. Otte, and J. K. S. Ching, 2004: Simulation of meteo-rological fields within and above urban and rural canopies with a mesoscale model (MM5). *Bound-Layer. Meteorol.*, 113 (1), 111–158.

Fernando, H. J. S., D. Zajic, S. D. Sabatino, R. Dimitrova, B. Hedquist, and A. Dallman, 2010: Flow, turbulence, and pollutant dispersion in urban atmospheres. *Physics of Fluids*, 22(5), 051 301.

Fiveland, W. A., 1984: Discrete-ordinates solutions of the radiative transport equation for rectangular enclosure. *J. Heat Tran.*, 106 (4), 699–706.

Fortuniak, K., B. Offerle, and C. S. B. Grimmond, 2005: Application of a slab surface energy balance model to determine surface parameters for urban areas. *Lund eRep. Phys. Geog.*, 5, 90–91.

Franke, J., A. Hellsten, H. Schlünzen, and B. Carissimo, 2007: Best Practice Guideline for the CFD simulation of flows in the urban environment: COST Action 732: Quality Assurance and Improvement of Microscale Meteorological Models. *Tech. Rep.* ISBN 3-00-018312-4, COST Office Brussels.

Gastellu-Etchegorry, J. P., 2008: 3D modeling of satellite spectral images, radiation budget and energy budget of urban landscapes. *Meteor. Atmos. Phys.*, 102 (3-4), 187–207.

Gastellu-Etchegorry, J. P., E. Martin, and F. Gascon, 2004: Dart: A 3-D model for simulating satellite images and surface radiation budget. *Int. J. of Remote Sens.*, 25, 75–96.

Grell, G., J. Dudhia, and D. R. Stauffer, 1994: *A Description of the Fifth-Generation Penn State/NCAR Mesoscale Model (MM5)*. NCAR/TN-398+STR, 138 pp.

Grimmond, C. S. B. and T. R. Oke, 1999a: Aerodynamic properties of urban areas derived from analysis of surface form. *J. Appl. Meteor.*, 38 (9), 1262–1292.

Grimmond, C. S. B. and T. R. Oke, 1999b: Heat storage in urban areas: Local-scale Observations and Evaluation of a Simple model. *J. Appl. Meteor.*, 38 (7), 922–940.

Grimmond, C. S. B. and T. R. Oke, 2002: Turbulent heat fluxes in urban areas: Observations & local-scale urban meteoro-logical parameterization scheme (LUMPS). *J. Appl. Meteor.*, 41 (7), 792–810.

Grimmond, C. S. B., et al., 2010: The International Urban Energy Balance Models Comparison Project: First Results from Phase 1. *J. Appl. Meteor. Climatol*, 49 (6), 1268–1292.

Grimmond, C. S. B., et al., 2011: Initial results from Phase 2 of the international urban energy balance model comparison. *Int. J. Climatol.*, 31 (2), 244–272.

Hamdi, R. and V. Masson, 2008: Inclusion of a drag approach in the Town Energy Balance (TEB) scheme: offline 1-D evaluation in a street canyon. *J. Appl. Meteor. Climatol*, 47 (10), 2627–2644.

Hanna, S. R., S. Tehranian, B. Carissimo, R. W. Macdonald, and R. Lohner, 2002: Comparisons of model simulations with observations of mean flow and turbulence within simple obstacle arrays. *Atmos. Environ.*, 36 (32), 5067–5079.

Hanna, S. R., et al., 2006: Detailed simulations of atmospheric flow and dispersion in downtown manhattan: An application of five computational fluid dynamics model. *Bull. Amer. Meteor. Soc.*, 87, 1713–1726.

Haoues-Jouve, S., et al., 2015: Meeting environmental quality requirements at neighbourhood scale: an original transdisciplinary approach allying human and physical sciences (The EUREQUA project). *Proceedings of the Ninth International Conference on Urban Climate, ICUC-9*, Toulouse, France, (July 20-24, 2015).

Harman, I. N. and S. E. Belcher, 2004: Radiative exchange in an urban street canyon. *Bound.-Layer Meteor.*, 110 (2), 301–316.

Harman, I. N. and S. E. Belcher, 2006: The surface energy balance and boundary layer over urban street canyons. *Quart. J. Roy. Meteor. Soc.*, 132 (621), 2749–2768.

Hémon, D. and E. Jougla, 2003: Estimation de la surmortalité et principales caractéristiques épidémiologiques. *Technical report*, [in French], INSERM.

Hénon, A., 2008: Températures mesurées, modélisées, et observées par télédétection infrarouge, dans la canopée urbaine: modélisation aéro-thermo-radiatif des flux de chaleur urbains. *Ph.D. thesis*, École Centrale de Nantes, 253 pp., [in French].

Hénon, A., P. G. Mestayer, J. P. Lagouarde, and J. A. Voogt, 2012: An urban neighborhood temperature and energy study from the CAPITOUL experiment with the SOLENE model. *Theor. Appl. Climatol.*, 110 (1-2), 177–196.

Holmes, N. and L. Morawska, 2006: A review of dispersion modelling and its application to the dispersion of particles: An overview of different dispersion models available. *Atmos. Environ.*, 40 (30), 5902–5928.

Howard, L., 1820: *The climate of London, deduced from Meteorological observations, made at different places in the neighbourhood of the metropolis. Vol 1.* Harvey and Darton, 358 pp.

Idczak, M., D. Groleau, P. Mestayer, J.-M. Rosant, and J.-F. Sini, 2010: An application of the thermo-radiative model SOLENE for the evaluation of street canyon energy balance. *Build. Environ.*, 45 (5), 1262–1275.

Idczak, M., P. Mestayer, J.-M. Rosant, J.-F. Sini, and M. Violleau, 2007: Micrometeorological measurements in a street canyon during the joint atreus-picada experiment. *Bound.-Layer Meteor.*, 124 (1), 25–41.

Jesionek, K. and M. Bruse, 2003: Impacts of vegetation on the microclimate: modeling standardized building structures with different greening levels. *Fifth Int. Conf. on Urban Climate*, Lodz, Poland (September 1-5, 2003).

Johnson, G. T., T. R. Oke, T. J. Lyons, D. G. Steyn, I. D. Watson, and J. A. Voogt, 1991: Simulation of surface urban heat islands under 'ideal' conditions at night. Part I: Theory and tests against field data. *Bound.-Layer Meteor.*, 56 (3), 275–294.

Kanda, M., , R. Moriwaki, and F. Kasamatsu, 2004: Large-eddy simulation of turbulent organized structures within and above explicitly resolved cube arrays. *Bound.-Layer Meteor.*, 112 (2), 343–368.

Kanda, M., 2006: Progress in the scale modeling of urban climate: Review. *Theor. Appl. Climatol.*, 84 (1-3), 23–33.

Kanda, M., T. Kawai, M. Kanega, R. Moriwaki, K. Narita, and A. Hagishima, 2005: A simple energy balance model for regular building arrays. *Bound.-Layer Meteor.*, 116 (3), 423–443.

Kato, M. and B. Launder, 1993: The modelling of turbulent flow around stationary and vibrating square cylinders. *Ninth symposium on turbulent shear flows*, Kyoto, Japan (August 16-18, 1993).

Kawai, T., M. Kanda, K. Narita, and A. Hagishima, 2007: Validation of a numerical model for urban energy-exchange using outdoor scale-model measurements. *Int. J. Climatol.*, 27 (14), 1931–1942.

Kawamoto, Y. and R. Ooka, 2009: Accuracy validation of urban climate analysis model using MM5 incorporating a multi-layer urban canopy model. *Seventh International Conference on Urban Climate, ICUC-7*, Yokohama, Japan, (June 28 - July 3, 2009) Extended abstract: A10-1.

Kim, J. and J. Baik, 2001: Urban street-canyon flows with bottom heating. *Atmos. Environ.*, 35 (20), 3395–3404.

Knaus, H., R. Schneider, X. Han, J. Ströjle, U. Schnell, and K. R. G. Hein, 1997: Comparison of different radiative heat transfer models and their applicability in coal-fired utility boiler simulations. *Fourth Int. Conf. on Technologies and Combustion for a Clean Environment*, Lisbon, Portugal (July 7-10, 1997).

Kondo, H., Y. Genchi, Y. Kikegawa, Y. Ohashi, H. Yoshikado, and H. Komiyama, 2005: Development of a Multi-Layer Urban Canopy Model for the Analysis of Energy Consumption in a Big City: Structure of the Urban Canopy Model and its Basic Performance. *Bound.-Layer Meteor.*, 116 (3), 395–421.

Kovar-Panskus, A., L. Moulinneuf, E. Savory, A. Abdelqari, J. F. Sini, J. M. Rosant, A. Robins, and N. Toy, 2002: A wind tunnel investigation of the influence of solar-induced wall-heating on the flow regime within a simulated urban street canyon. *Water, Air, & Soil Pollution*, 2 (5-6), 555–571.

Krayenhoff, E. S. and J. A. Voogt, 2007: A microscale three-dimensional urban energy balance model for studying surface temperatures. *Bound.-Layer Meteor.*, 123 (3), 433–461.

Kusaka, H., H. Kondo, Y. Kikegawa, and F. Kimura, 2001: A simple single-layer urban canopy model for atmospheric models: comparison with multi-layer and slab models. *Bound.-Layer Meteor.*, 101 (3), 329–358.

Lagouarde, J. P., A. Hénon, B. Kurz, P. Moreau, M. Irvine, J. Voogt, and P. Mestayer, 2010: Modelling daytime thermal infrared directional anisotropy over Toulouse city centre. *Remote Sens. Environ.*, 114 (1), 87–105.

Launder, B. E., G. J. Reece, and W. Rodi, 1975: Progress in the Development of a Reynolds-Stress Turbulent Closure. *J. Fluid Mech.*, 68 (3), 537–566.

Launder, B. E. and D. B. Spalding, 1974: The numerical computation of turbulent flow. *Comp. Meth. Appl. Mech. Eng.*, 3, 269–289.

Lawson, T. V. and A. D. Penwarden, 1975: The effects of wind on people in the vicinity of buildings. *Fourth Int. Conf. on Wind Effects on Buildings and Structures*, Heathrow, Cambridge University Press, (pp. 605-622).

Lee, S.-H. and S.-U. Park, 2004: A vegetated urban canopy model for meteorological and environmental modelling. *Bound.-Layer Meteor.*, 126 (1), 73–102.

Lemonsu, A., C. S. B. Grimmond, and V. Masson, 2004: Modelling the surface energy balance of an old Mediterranean city core. *J. Appl. Meteor.*, 43 (2), 312–327.

Liu, J., H. M. Shang, and Y. S. Chen, 2000: Development of an unstructured radiation model applicable for two-dimensional planar, axisymmetric, and three-dimensional geometries. *J. Quant. Spectrosc. Radiat. Transfer*, 66 (1), 17–33.

Loomans, M. G. L. C., M. van Houdt, A. D. Lemaire, and J. L. M. Hensen, 2008: Performance assessment of an operating theatre design using CFD simulation and tracer gas measurements. *Indoor and Built Environment*, 17 (4), 299–312.

Louis, J., 1979: A parametric model of vertical eddy fluxes in the atmosphere. *Bound.-Layer Meteor.*, 17 (2), 187–202.

Louka, P., G. Vachon, J. F. Sini, P. G. Mestayer, and J. M. Rosant, 2001: Thermal effects on the airflow in a street canyon –Nantes 99 Experimental Results and Model Simulation. *Water, Air, and Soil Pollut.*, 2 (5-6), 351–364.

Macdonald, R. W., 2000: Modelling the mean velocity profile in the urban canopy layer. *Bound.-Layer Meteor.*, 197 (1), 25–45.

Martilli, A., A. Clappier, and M. W. Rotach, 2002: An urban surface exchange parameterisation for mesoscale models. *Bound.-Layer Meteor.*, 104 (2), 261–304.

Masson, V., 2000: A physically based scheme for the urban energy budget in atmospheric models. *Bound.-Layer Meteor.*, 94 (3), 357–397.

Masson, V., 2006: Urban surface modeling and the meso-scale impact of cities. *Theor. Appl. Climatol.*, 84 (1-3), 35–45.

Masson, V., C. S. B. Grimmond, and T. R. Oke, 2002: Evaluation of the Town Energy Balance (TEB) Scheme with Direct Measurements from Dry Districts in Two cities. *J. Appl. Meteor.*, 41 (10), 1011–1026.

Masson, V., et al., 2008: The Canopy and Aerosol Particles Interactions in TOulouse Urban Layer (CAPITOUL) experiment. *Meteor. Atmos. Phys.*, 102 (3-4), 135–157.

Maxwell, E. L., 1998: METSTAT-the solar radiation model used in the production of the National Solar Radiation Data Base (NSRDB). *Solar Energy*, 62 (4), 263–279.

McCarthy, M. P., M. J. Best, and R. A. Betts, 2010: Climate change in cities due to global warming and urban effects. *Geophys. Res. Lett.*, 37, L09705, doi:10.1029/2010GL042845.

Miguet, F. and D. Groleau, 2002: A daylight simulation tool for urban and architectural spaces - application to transmitted direct and diffuse light through glazing. *Build. Environ.*, 37 (8), 833–843.

Milliez, M., 2006: Modélisation micro-météorologique en milieu urbain: dispersion des polluants et prise en compte des effets radiatifs. *Ph.D. thesis*, École des Ponts ParisTech, 228 pp., [in French, available on line at http://cerea.enpc.fr/fr/theses.html].

Milliez, M. and B. Carissimo, 2007: Numerical simulations of pollutant dispersion in an idealized urban area, for different meteorological conditions. *Bound.-Layer Meteor.*, 122 (2), 321–342.

Milliez, M. and B. Carissimo, 2008: CFD modelling of concentration fluctuations in an idealized urban area. *Bound.-Layer Meteor.*, 127 (2), 241–259.

Mills, G., 1997: An Urban Canopy-Layer Climate Model. *Theor. Appl. Climatol.*, 57 (3-4), 229–244.

Moin, P. and K. Mahesh, 1998: Direct numerical simulation: A tool in turbulence research. *Annu. Rev. Fluid Mech.*, 30, 539–578.

Monsi, M. and T. Saeki, 1953: Über den Lichtfaktor in den Pflanzengesellschaften und seine Bedeutung für die Stoffproduktion. *Japanese Journal of Botany*, 14, 22–52.

Monteith, J. L. and M. H. Unsworth, 1990: *Principles of environment physics*. 2d ed., Edward Arnold, 291 pp.

Moscicki, M. A., 2007: A Comparison of Modeled and Observed Surface Temperatures in Toulouse, France. *Ph.D. thesis*, The University of Western Ontario, 170 pp.

Musson-Genon, L., 1993: Paramétrisation du rayonnement infra-rouge dans le code atmosphérique mercure (version 3.1). *Rapport interne EDF*, [in French], EDF R&D, Fluid Mechanics and Heat Transfer Department.

Musson-Genon, L., 1994: Paramétrisation du rayonnement solaire dans le code atmosphérique mercure (version 3.1). *Rapport interne EDF*, [in French], EDF R&D, Fluid Mechanics and Heat Transfer Department.

Musson-Genon, L., E. Dupont, and D. Wendum, 2007: Reconstruction of the Surface-Layer Vertical Structure from Measurements of Wind, Temperature and Humidity at Two Levels. *Bound.-Layer Meteor.*, 124 (2), 235–250.

Neophytou, M., et al., 2011: Towards optimization of urban planning and architectural parameters for energy use minimization in Mediterranean cities. *World Renewable Energy Congress 2011*, Linköping, Sweden (May 8-11, 2011).

Niachou, K., I. Livada, and M. Santamouris, 2008: Experimental study of temperature and airflow distribution inside an urban canyon during hot summer weather conditions. Part II: Airflow analysis. *Build. Environ.*, 43 (8), 1393–1403.

NIWA, 2004: Good Practice Guide for Atmospheric Dispersion Modelling. *Technical report*, National Institute of Water and Atmospheric Research, New Zealand.

Offerle, B., C. S. B. Grimmond, and T. R. Oke, 2003: Parameterization of net all-wave radiation for urban areas. *J. Appl. Meteor. Climatol.*, 42 (8), 1157–1173.

Oke, T. R., 1987: *Boundary Layer Climates*. 2d ed., Routledge, 435 pp.

Oke, T. R., R. Spronken-Simith, E. Jáuregui, and C. S. B. Grimmond, 1999: The energy balance of central Mexico City during the dry season. *Atmos. Environ.*, 33 (24-25), 3919–3930.

Oleson, K. W., G. B. Bonan, J. Feddema, M. Vertenstein, and C. S. B. Grimmond, 2008: An urban parameterization for a global climate model: 1. Formulation & evaluation for two cities. *J. Appl. Meteor. Climatol.*, 47 (4), 1038–1060.

Onomura, S., H. Takimoto, and M. Kanda, 2009: Influence of a heated wall on urban canopy flow using PIV measurements. *Seventh Int. Conf. on Urban Climate*, Yokohama, Japan (June 29 - July 3, 2009).

Orlanski, I., 1975: A rational subdivision of scales for atmospheric processes. *B. Am. Meteorol. Soc.*, 56, 527–530.

Pasquill, F. and F. Smith, 1983: *Atmospheric diffusion*. 3d ed., New York: Halsted Press, John Wiley and Sons.

Péniguel, C. and I. Rupp, 2008: SYRTHES 3.4 - Manuel Théorique. *Technical report*, [in French], EDF R&D, Fluid Mechanics and Heat Transfer Department.

Perez, R., R. Seals, and J. Michalsky, 1993: An all-weather model for sky luminance distribution. *Sol. Energ.*, 50, 235–245.

Pigeon, G., D. Legain, P. Durand, and V. Masson, 2007: Anthropogenic heat release in an old European agglomeration (Toulouse, France). *Int. J. Climatol.*, 27 (14), 1969–1981.

Pigeon, G., M. A. Moscicki, J. A. Voogt, and V. Masson, 2008: Simulation of fall and winter surfac energy balance over a dense urban area using the TEB scheme. *Meteorol. Atmos. Phys.*, 102 (3-4), 159–171.

Pope, S. B., 2000: Turbulent flows. *Cambridge University Press*, 125, 1361–1362.

Prandtl, L., 1925: Bericht über untersuchungen zur ausgebildeten turbulenz, Z. angew. *Math, Meth.*, 5, 136–139.

Prata, A. J., 1996: A new long-wave formula for estimating downward clear-sky radiation at the surface. *Q. J. R. Meteorol. Soc.*, 122, 1127–1151.

Qu, Y., M. Milliez, L. Musson-Genon, and B. Carissimo, 2011a: Impacts on airflow of thermal transfers taking into account 3D atmospheric radiation in urban areas. *13th International Conference on Wind Engineering (ICWE 13)*, Amsterdam, Netherlands (July 10-15, 2011).

Qu, Y., M. Milliez, L. Musson-Genon, and B. Carissimo, 2011b: Micrometeorological modeling of radiative and convective effects with a building-resolving code. *J. Appl. Meteor. Climatol.*, 50(8), 1713–1724.

Qu, Y., M. Milliez, L. Musson-Genon, and B. Carissimo, 2011c: Modeling of the urban energy balance taking into account fluid mechanics with meteorology in an idealized urban area. *Newsletter d'EURASAP 72*, April 2011.

Qu, Y., M. Milliez, L. Musson-Genon, and B. Carissimo, 2012a: Development of a building resolving atmospheric CFD code taking into account atmospheric radiation in complex geometries. *Air Pollution Modeling and its Application XXI, NATO Science for Peace and Security Series C: Environmental Security*, 4, 75–80.

Qu, Y., M. Milliez, L. Musson-Genon, and B. Carissimo, 2012b: Numerical study of the thermal effects of buildings on low-speed airflow taking into account 3d atmospheric radiation. *J. Wind Eng. Ind. Aerodyn.*, 104–106, 474–483.

Qu, Y., M. Milliez, L. Musson-Genon, and B. Carissimo, 2013: Numerical simulations of pollutant dispersion in the centre of a European city for different thermal transfer conditions. *15th international conferences on Harmonisation within Atmospheric Dispersion Modelling for Regulatory Purposes (HARMO15)*, Madrid, Spain (May 6-9, 2013).

Qu, Y., M. Milliez, L. Musson-Genon, and B. Carissimo, 2014: 3D Radiative and Convective Modeling of Urban Environment: An Example for the City Center of Toulouse. *Air Pollution Modeling and its Application XXII, NATO Science for Peace and Security Series C: Environmental Security*, 727–731.

Randerson, D., 1976: An overview of regional-scale numerical models. *Bull. Am. Met. Soc.*, 57, 797–804.

Richards, K., M. Schatzmann, and B. Leitl, 2006: Wind tunnel experiments modelling the thermal effects within the vicinity of a single block building with leeward wall heating. *J. Wind Eng. Ind. Aerodyn.*, 94 (8), 621–636.

Ringenbach, N., 2004: Bilan radiatif et flux de chaleur en climatologie urbaine: mesures, modélisation et validation sur strasbourg. *Ph.D. thesis*, Université Louis Pasteur Strasbourg I, 180 pp., [in French].

Rotach, M. W., 1995: Profiles of turbulence statistics in and above an urban street canyon. *Atmos. Environ.*, 29, 1473–1486.

Ryu, Y.-H., , J.-J. Baik, and S.-H. Lee, 2009: A new single-layer urban canopy model for use in mesoscale atmospheric models. *Proceedings of the Seventh International Conference on Urban Climate, ICUC-7*, Yokohama, Japan, (June 28 - July 3, 2009) Extended abstract: A10-1.

Salamanca, F., A. Krpo, A. Martilli, and A. Clappier, 2009: A new building energy model coupled with an urban canopy parameterization for urban climate simulations - Part I: formulation, verification, and sensitivity analysis of the model. *Theor. Appl. Climatol.*, 99 (3-4), 331–344.

Santiago, J. L., A. Dejoan, A. Martilli, F. Martin, and A. Pinelli, 2010: Comparison Between Large-Eddy Simulation and Reynolds-Averaged Navier-Stokes Computations for the MUST Field Experiment. Part I: Study of the Flow for an Incident Wind Directed Perpendicularly to the Front Array of Containers. *Bound.-Layer Meteor.*, 135 (1), 109–132.

Schlünzen, K. H., D. Grawea, S. I. Bohnenstengelb, I. Schlüterc, and R. Koppmannd, 2011: Joint modelling of obstacle induced and mesoscale changes–Current limits and challenges. *J. Wind Eng. Ind. Aerodyn.*, 99 (4), 217–225.

Shashua-Bar, L. and M. E. Hoffman, 2004: Quantitative evaluation of passive cooling of the UCL microclimate in hot regions in summer. *Build. Environ.*, 39 (9), 1087–1099.

Shirasawa, T., R. Yoshie, H. Tanaka, T. Kobayashi, A. Mochida, and Y. Endo, 2008: Cross comparison of CFD results of gas diffusion in weak wind region behind a high-rise building. *Proceedings of the 4th International Conference on Advances in Wind and Structures (AWAS'08)*, Jeju, Korea, (May 2008) 1038-1050.

Sini, J. F., S. Anquetin, and P. Mestayer, 1996: Pollutant Dispersion and Thermal Effects in Urban Street Canyons. *Atmos. Environ.*, 30 (15), 2659–2677.

Soux, A., J. A. Voogt, and T. R. Oke, 2004: A Model to Calculate What a Remote Sensor 'Sees' of an Urban Surface. *Bound.-Layer Meteorol.*, 111 (1), 109–132.

Stephens, G. L., 1984: The parameterization of radiation for numerical weather prediction and climate models. *Mon. Weather. Rev.*, 112, 826–867.

Stull, R. B., 1988: *An Introduction To Boundary Layer Meteorology*. Kluwer Academic Publisher, Dordrecht, The Netherlands, 670 pp.

Tominaga, Y., A. Mochida, R. Yoshie, H. Kataoka, T. Nozu, M. Yoshikawa, and T. Shirasawa, 2008: AIJ guidelines for practical applications of CFD to pedestrian wind environment around buildings. *J. Wind Eng. Ind. Aerodyn.*, 96 (10-11), 1749–1761.

Truelove, J. S., 1987: Discrete-ordinates solutions of the radiative transport equation. *J. Heat Tran.*, 109, 1048–1051.

Uehara, K., S. Murakami, S. Oikawa, and S. Wakamatsu, 2000: Wind tunnel experiments on how thermal stratification affects flow in and above urban street canyons. *Atmos. Environ.*, 34 (10), 1553–1562.

Washington, W. M., 1972: Numerical climatic-change experiments: The effect of mans production of thermal energ. *J. Appl. Meteorol*, 11 (5), 768–772.

WBCSD, 2009: *Energy Efficiency in Buildings*. WBCSD news, World Business Council for Sustainable Development.

WHO, 2006: Air quality guidelines. Global update 2005. particulate matter, ozone, nitrogen dioxide and sulfur dioxide. *Technical report*, The World Health Organization.

Xie, X., C. H. Liu, and D. Y. C. Leung, 2007: Impact of building facades and ground heating on wind flow and pollutant transport in street canyons. *Atmos. Environ.*, 41 (39), 9030–9049.

Xie, Z.-T., O. Coceal, and I. P. Castro, 2008: Large-eddy simulation flows over random urban-like obstacles. *Bound.-Layer Meteor.*, 129 (1), 1–23.

Yakhot, V., S. A. Orszag, S. Thangam, T. B. Gatski, and C. G.Speziale, 1992: Development of turbulence models for shear flows by a double expansion technique. *Phys. Fluids*, 4, 1510–1520.

Yamada, T., 1982: A numerical study of turbulent airflow in and above a forest canopy. *J. Met. Soc. Japan*, 60, 439–454.

Yang, K., G. W. Huang, and N. Tamai, 2001: A hybrid model for estimating global solar radiation. *Solar Energy*, 70 (1), 13–22.

Yee, E. and C. A. Biltoft, 2004: Concentration fluctuations measurements in a plume dispersing through a regular array of obstacles. *Bound.-Layer Meteorol.*, 111 (3), 363–415.

Zaïdi, H., E. Dupont, M. Milliez, B. Carissimo, and L. Musson-Genon, 2013: Numerical simulations of the microscale heterogeneities of turbulence observed on a complex site. *Bound.-Layer Meteorol.*, 147, 237–259.

AUTHOR'S CONTACT INFORMATION

Dr. Yongfeng Qu
GTT (Gaztransport & Technigaz) SA
1, Route de Versailles
78470 Saint-Rémy-lès-Chevreuse, France
Email: yongfeng.qu@cerea.enpc.fr

INDEX

D

E

F

G

H

I

J

K

L

M

N

O

P

Q

R

U

V

W

Y